www.KnowledgePublications.com

SAVE HARMLESS AGREEMENT

Because use of the information, instructions and materials discussed and shown in this book, document, electronic publication or other form of media is beyond our control, the purchaser or user agrees, without reservation to save Knowledge Publications Corporation, its agents, distributors, resellers, consultants, promoters, advisers or employees harmless from any and all claims, demands, actions, debts, liabilities, judgments, costs and attorney's fees arising out of, claimed on account of, or in any manner predicated upon loss of or damage to property, and injuries to or the death of any and all persons whatsoever, occurring in connection with or in any way incidental to or arising out of the purchase, sale, viewing, transfer of information and/or use of any and all property or information or in any manner caused or contributed to by the purchaser or the user or the viewer, their agents, servants, pets or employees, while in, upon, or about the sale or viewing or transfer of knowledge or use site on which the property is sold, offered for sale, or where the property or materials are used or viewed, discussed, communicated, or while going to or departing from such areas.

Laboratory work, scientific experiment, working with hydrogen, high temperatures, combustion gases as well as general chemistry with acids, bases and reactions and/or pressure vessels can be EXTREMELY DANGEROUS to use and possess or to be in the general vicinity of. To experiment with such methods and materials should be done ONLY by qualified and knowledgeable persons well versed and equipped to fabricate, handle, use and or store such materials. Inexperienced persons should first enlist the help of an experienced chemist, scientist or engineer before any activity thereof with such chemicals, methods and knowledge discussed in this media and other material distributed by KnowledgePublications Corporation or its agents. Be sure you know the laws, regulations and codes, local, county, state and federal, regarding the experimentation, construction and or use and storage of any equipment and or chemicals BEFORE you start. Safety must be practiced at all times. Users accept full responsibility and any and all liabilities associated in any way with the purchase and or use and viewing and communications of knowledge, information, methods and materials in this media.

better farming series 31

biogas

what it is
how it is made
how to use it

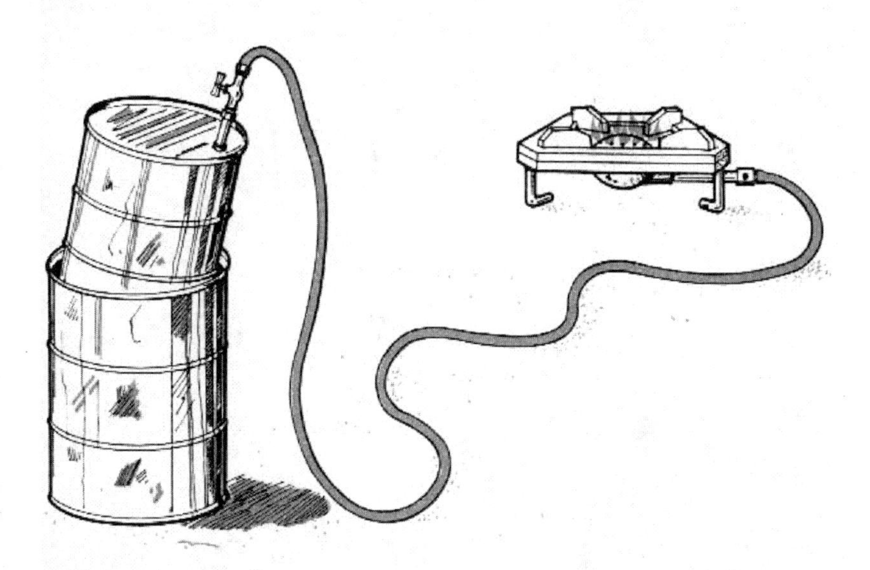

FOOD AND AGRICULTURE ORGANIZATION OF THE UNITED NATIONS

Biogas

What it is
How it is made
How to use it

FOOD AND AGRICULTURE ORGANIZATION OF THE UNITED NATIONS
Rome 1984

First published 1984

Translated, modified, printed and made available on the Internet and elsewhere by the People of Africa Biogas Centre. This document may be photocopied and distributed for the purposes of instruction, education and the assistance of the many who need better and clean energy.

-People of Africa Biogas Centre 1996
Democratic Republic of the Congo

ISBN
978-1-60322-031-6

PREFACE

The first twenty-six volumes in FAO's Better Farming Series were based on the **Cours d'apprentissage agricole** prepared in the Ivory Coast by the **Institut africain de développement économique et social** for use by extension workers. Later volumes, beginning with No. 27, have been prepared by FAO for use in agricultural development at the farm and family level. The approach has deliberately been a general one, the intention being to constitute basic prototype outlines to be modified or expanded in each area according to local conditions of agriculture.

Many of the booklets deal with specific crops and techniques, while others are intended to give the farmer more general information which can help him to understand **why** he does what he does, so that he will be able to do it better.

Adaptations of the series, or of individual volumes in it, have been published in Amharic, Arabic, Bengali, Creole, Hindi, Igala, Indonesian, Kiswahili, Malagasy, SiSwati and Turkish, an indication of the success and usefulness of this series.

Requests for permission to issue this manual in other languages and to adapt it according to local climatic and ecological conditions are welcomed. They should be addressed to the Director, Publications Division, Food and Agriculture Organization of the United Nations, Via delle Terme di Caracalla, 00100 Rome, Italy.

OUTLINE OF COURSE

INTRODUCTION

1. Farmers and their families
   always look for ways
   to make their lives better.

2. One way farm families
   can make their lives better
   is to make their own fuel gas
   which they can use for cooking.

3. Today many farmers
   are making fuel gas at home.
   They make it from animal manure
   or from plant materials
   or from a mixture of both.

4. Fuel gas made in this way
   has a lot of methane in it.
   Methane burns very well.

5. Where you live
methane gas may be called
by a different name.
One of the most common names
for this kind of gas,
when it is made at home,
is **biogas.**
We will use the name **biogas**
in this booklet.

Biogas

6. If you make your own biogas
you will not have to use
so much of the more expensive fuels
such as kerosene and charcoal
or firewood, which may be hard to find
where you live.

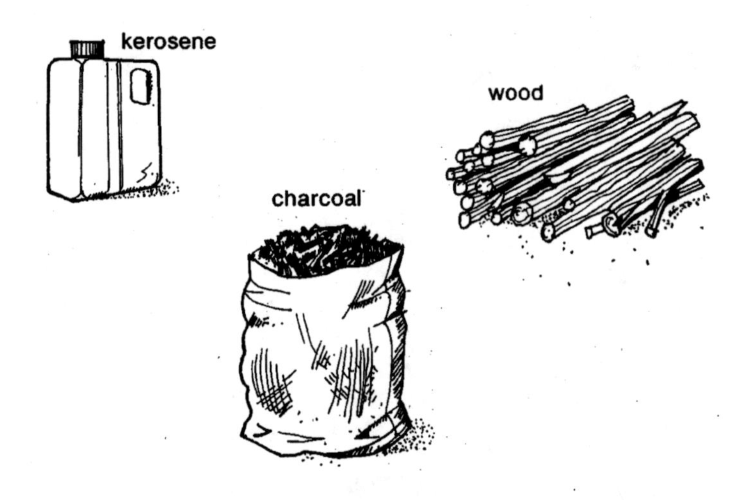

7. Using biogas can help you
to save time and work
when you cook your meals.

8. You can use the time you save
to do other things
around your home and farm,
such as care for a bigger garden

or work at a money-making home craft.

9. Biogas is a clean-burning fuel.
It does not give off smoke
as does charcoal or firewood.
By using biogas for cooking
you can keep your cooking area
and your food cleaner.

10. After all gas has been made
the material that is left
is a very rich fertilizer
that you can use on your fields.

11. This booklet was written
to help you to learn
some of the things that you need to know
before you begin to make biogas.
You will also learn
how to make your own biogas.

How is biogas made?

12. When animal manure
or plant materials rot
they give off gas.
You collect this gas as it is made
when you make biogas.

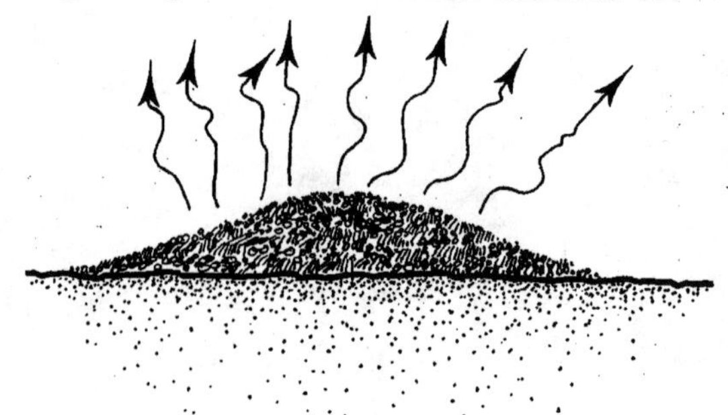

13. In this booklet
you will learn how to mix water
with animal manure or plant materials,
how to put this mixture into a container
where it will rot and give off gas,
and how to collect the gas
in another container which is airtight.
We will call these containers
the **biogas unit.**

14. It is not easy to build a biogas unit.
When you begin
you will have to spend a lot of time
and work very hard.
It may also cost you money.

15. You must be sure
that building a biogas unit
will be a good way
to use your time and money.

16. You will need a good place
to put your biogas unit.
Items 26 to 30 will tell you
where to put it.

17. If you live where it is too hot
or too cold,
your may find it hard
to keep a biogas unit working.

18. **Biogas is produced best
at a temperature
between 32 and 37°C.**
When the temperature
is below 15°C
almost no gas is made.
Items 31 to 34
will tell you some ways
which will help you
to keep your biogas unit
at the right temperature.

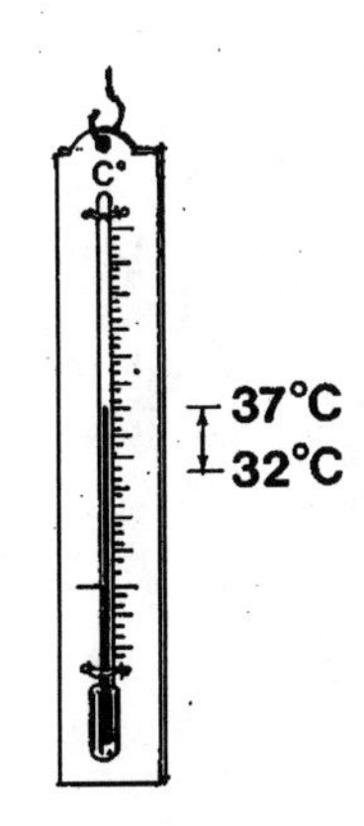

19. You will need oil drums, pipe, valves, a gas line and sealing materials to build a biogas unit.

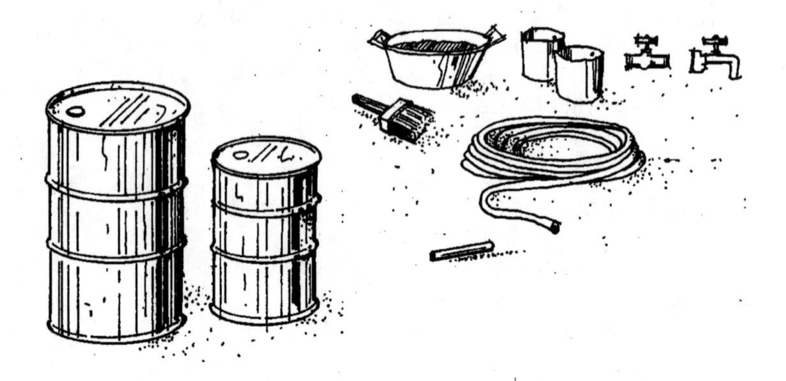

You will need a good supply of animal manure or plant material.

20. If there is a biogas unit near where you live, you should go to visit it. Talk to those who have built it and are running it to see how it works.

21. When you are thinking about building a biogas unit, you may be able to get advice from your extension officer.

How big should your biogas unit be?

22. Begin by building a small unit. Items 35 to 57 will tell you how. With a small biogas unit, you will need less animal manure and plant materials. A small unit will cost less to build and it will be easier to run.

23. When you have learned
how to run your biogas unit
and have made and used your own gas,
you may decide
that you need more gas.

24. You can get more gas
by building one or more biogas units
just like your first one.
Items 126 to 130 will tell you
how to run several biogas units together.

HOW TO BUILD A SMALL BIOGAS UNIT

25. You can build a small biogas unit from two oil drums.

You will need

- an oil drum of about 200 litres, to hold the waste
- an oil drum of about 120 litres, to collect the gas

200-litre drum

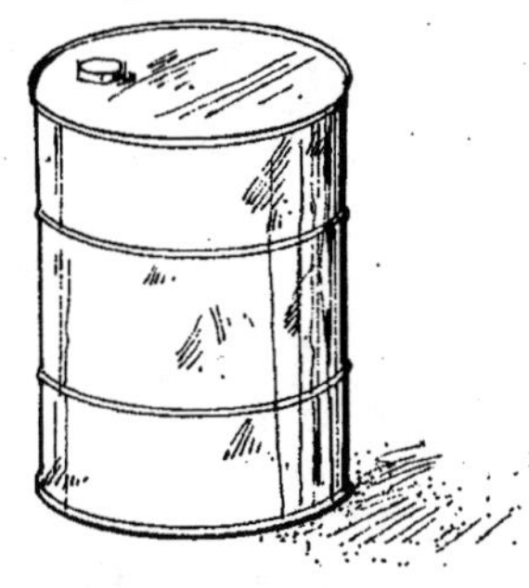

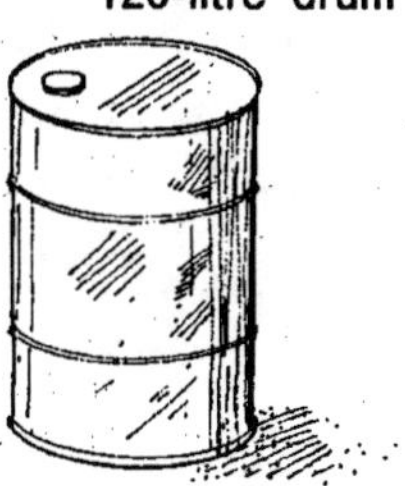

- a piece of pipe about 10 centimetres long and about 2 centimetres in diameter, for the gas outlet
- a valve to fit the gas outlet

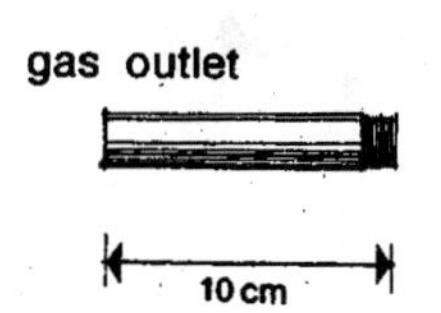

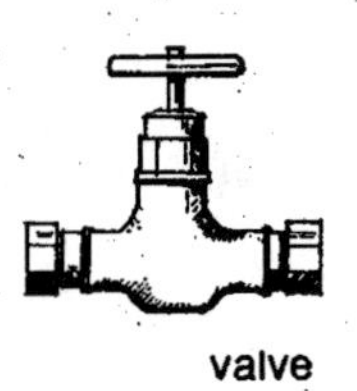

valve

- at least 10 metres
  of rubber or plastic tube
  about 2 centimetres in diameter,
  for the gas line

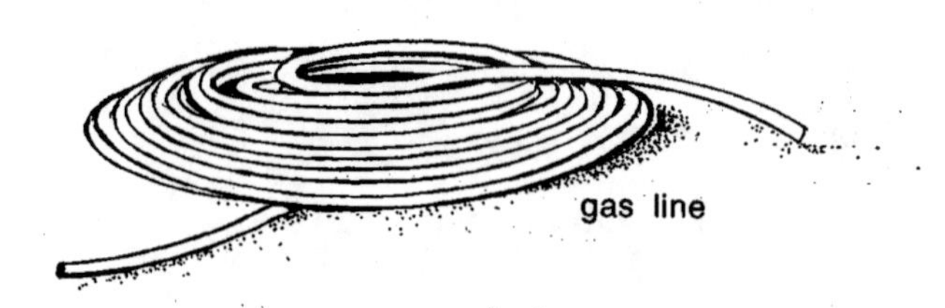

Where to put your biogas unit

26. Be careful
    not to put your biogas unit
    too close to your home
    or your cooking area
    or your water supply.

27. A biogas unit should be
    at least **10 metres** from your home
    so that when you put waste
    into your unit
    it will not be too close
    to where you and your family live
    and cook your meals.

28. Do not put your biogas unit
too far from where you cook
or you will need a long gas line.
Gas lines are hard to find
and may cost a lot of money.

29. If your gas line is moved
or damaged, it may leak
when gas is made.
If your gas line crosses a path,
bury it a little underground
to protect it
from being moved or damaged.

30. A biogas unit
should be at least **15 metres**
from your water supply,
so that the waste in your unit
will not make your water dirty
and unhealthy to drink or use.

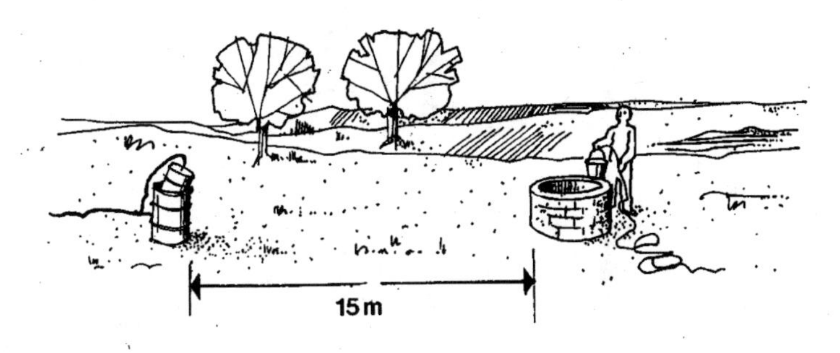

31. You have already been told
that you will get most gas
if the temperature of your unit
is between 32 and 37°C.

32. If you live in a very hot place,
put your unit out of the sun,
in the shade or under trees
to keep it from getting too hot.

33. If you live in a place
that is not very warm,
put your unit in the sun
to keep it warm.

34. If you live in a cold place,
put the unit underground
or cover it with earth or straw
to keep it warm.

Building the unit

35. The bottom part of the unit,
which holds the waste mixture,
is made from the bigger drum.
The top part of the unit,
which holds the gas,
is made from the smaller drum
which you put inside the bigger drum.

36. Most drums have a hole in the top.
You will not need a hole
in the top of the bigger drum
but you will need a hole
in the top of the smaller drum
for the gas outlet.

37. Cut out one end from each drum.
You can do this
using a hammer and metal chisel.
Cut the end of the bigger drum
that has a hole in it.
Cut the end of the smaller drum
that does not have a hole in it.

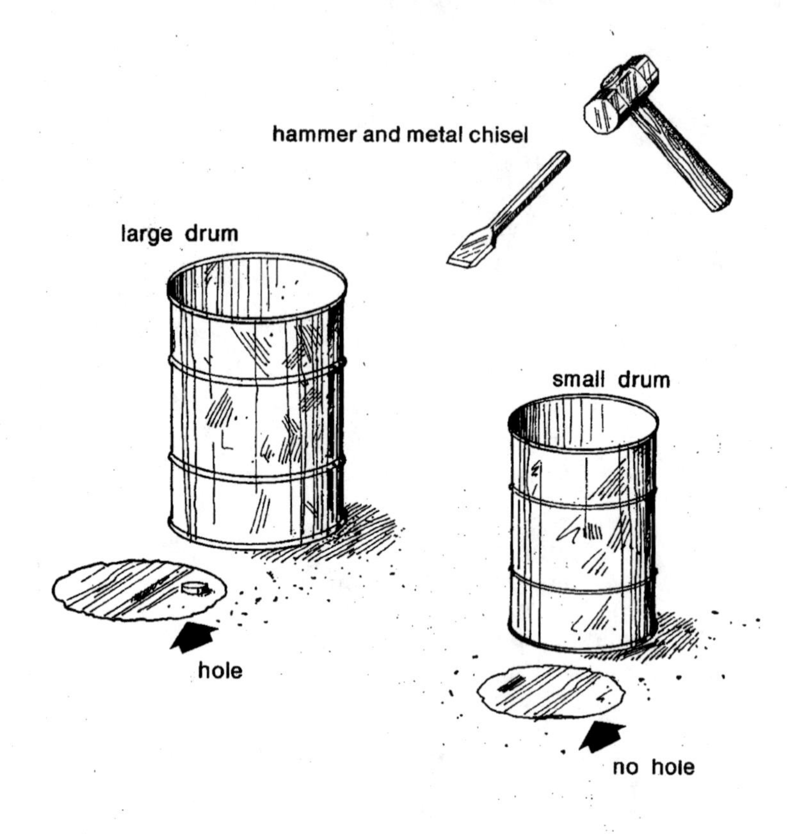

38. If the small drum
does not have a hole in the top,
you will have to cut one
(see Item 45).

39. Now clean both drums well inside and outside to remove oil and grease.

40. If either drum has a hole in the side, close it tightly. This can be done with a metal plug or by welding a piece of metal in the hole as shown in the drawings.

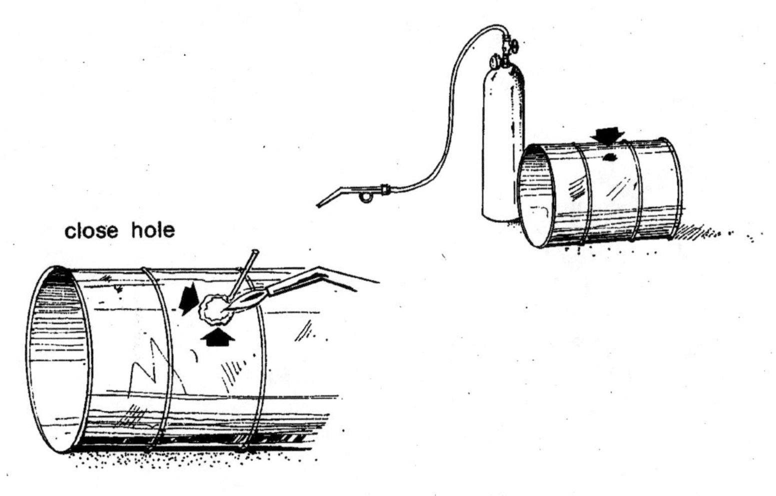

41. Now you are ready
to put the gas outlet
in the top of the small drum.

42. The gas outlet
is made from a short piece of pipe
about 10 centimetres long
and about 2 centimetres in diameter.

43. If there is a threaded hole
in the top of the small drum,
use a gas outlet
which is threaded on both ends
and screw it tightly into the hole.

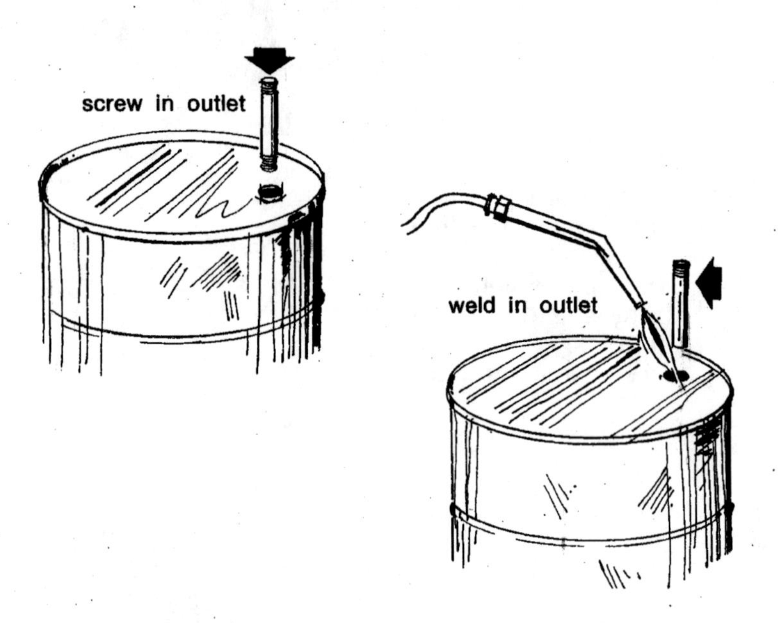

44. If the hole
in the top of the small drum
is not threaded,
use a gas outlet
which is threaded on one end
and weld it into the hole
with the threaded end up.

45. If there is no hole
in the top of the small drum,
cut one about 2 centimetres in diameter
and using a gas outlet
which is threaded on one end
weld it into the hole
with the threaded end up,
as you were told in item 44.

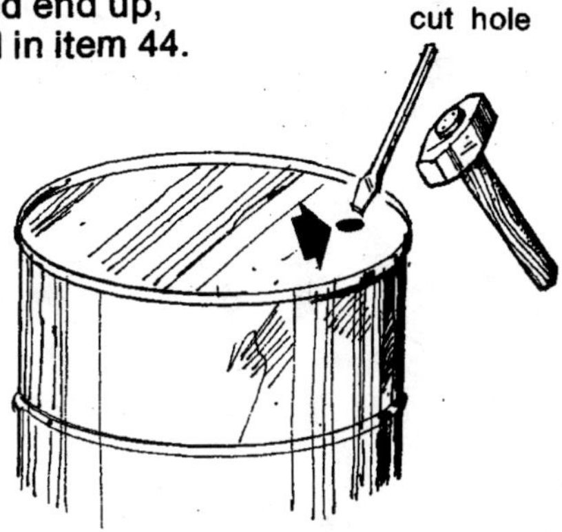

46. Now you are ready
to attach a valve
to the top of the gas outlet.
You can use valves
like the ones shown in the drawing.

47. The valve you use must be airtight
so that it will not leak gas
and you must be sure
to screw it tightly to the gas outlet.

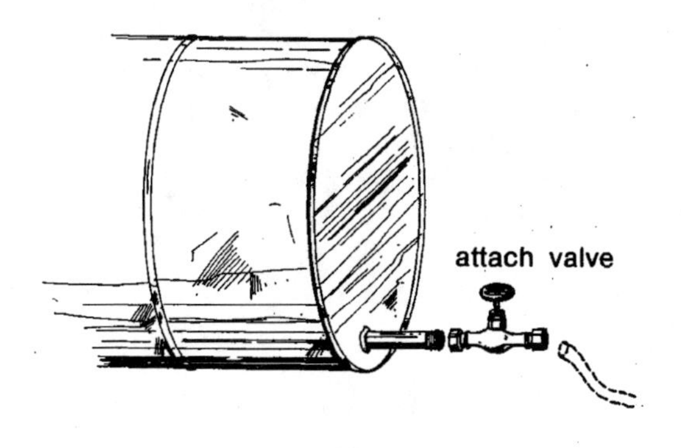

48. If you do not have a valve,
attach the rubber or plastic tube
you are using for the gas line
directly to the gas outlet.
To close the gas line,
you can fold it once
and clamp it shut
or you can fold it twice
and tie it tightly with cord
as shown in the drawings.

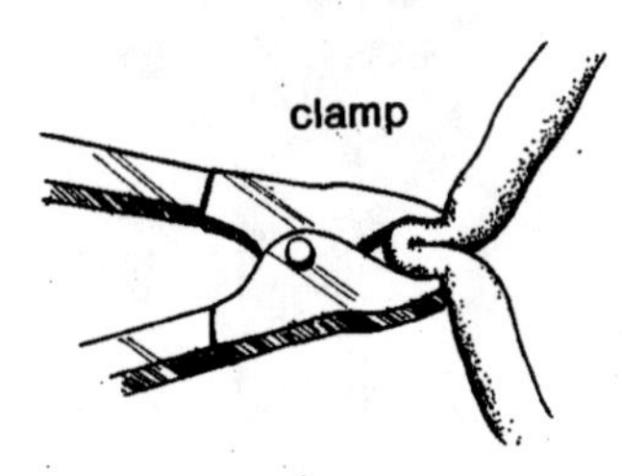

Testing for leaks

49. Now you are ready
to test the small drum for leaks.
To hold gas it must be airtight.

50. To check for leaks, close the valve
or if you have no valve
clamp or tie the gas line tightly
as you were told in item 48.

51. Turn the small drum over
and place it above the ground
on stones or pieces of wood,
but be careful not to damage
the gas outlet or the valve
or to loosen the clamp
or the cord on the gas line.
Now fill the small drum with water.

52. If you see water leaking from the drum, the gas outlet, the valve or the tied gas line, mark the place of each leak. Then empty out the water, being careful not to damage the gas outlet, and let the drum dry.

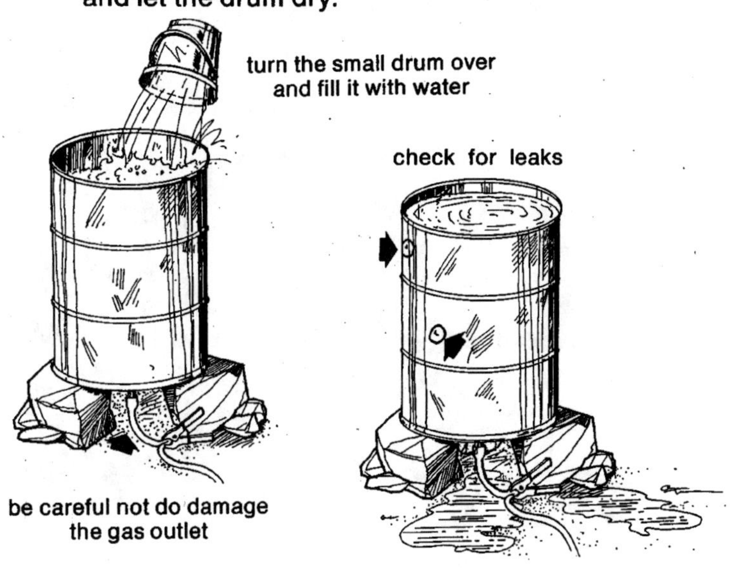

53. Seal the leaks by coating them with tar, mastic or paint on the inside and the outside of the drum.

54. If there are leaks around the gas outlet or valve, tighten the outlet or valve again and coat the joints with tar, mastic or paint.

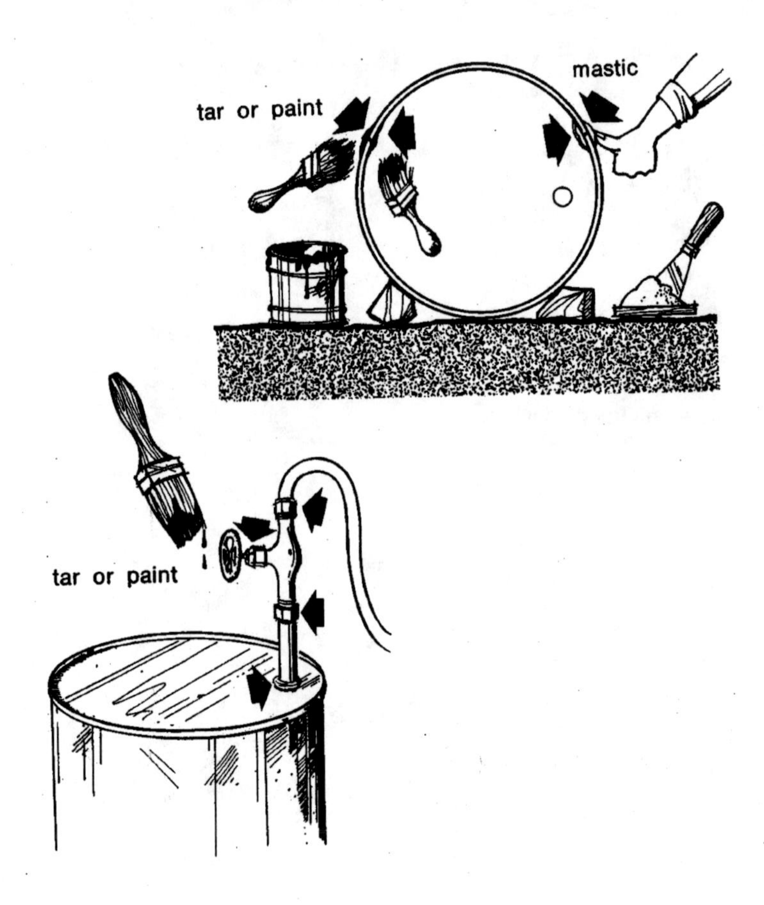

55. When the sealing is dry,
fill the drum with water again
to be sure that the leaks are well sealed.
If water still leaks,
start over again.

56. When the small drum is well sealed
and no longer leaks,
your biogas unit is ready to use.

57. **It is very important
to seal all leaks carefully.**

WASTE MATERIALS

What waste materials to use

58. Animal manure
from cows, pigs and chickens,
and crop and plant wastes
are good materials for making biogas.

59. You can use animal manure alone
or plant materials alone
or you can use both mixed together.

60. Straw which is mixed with manure,
which you may have
where you keep your pigs or chickens,
is usually a good mixture
of animal manure and plant material
for making biogas.
Be careful to chop it fine
before you use it.

61. When you first begin,
it is best to use only animal manure
or a mixture of animal manure
and very little plant material.
Later when you have learned more
about how your biogas unit works,
you can use more plant materials.

62. When you do begin to use plant materials,
remember that dry plant materials
must be chopped or shredded very fine
and fresh plant materials
must be left outside to rot
for 10 days or more
before you put them into a biogas unit.

How to use them

63. Whether you are going to use animal manure or plant materials in your biogas unit, you must mix them with water. Use one bucket of water with every bucket of animal manure or plant material.

64. Plant materials which are not mixed well may not make gas later. When you mix plant materials with water, they pack together.

65. If you are using plant materials, break them apart and stir them well so that they will be well mixed.

66. Mix the animal manure or the plant material with water until the waste mixture is easy to pour. The waste mixture will work best if it is like a thin paste.

Making a starter

67. About two months
before you are ready
to use your biogas unit
for the first time,
put 2 litres of animal manure
and 2 litres of water
in a bucket and mix well.
You can also add
some finely chopped plant material
such as grass.

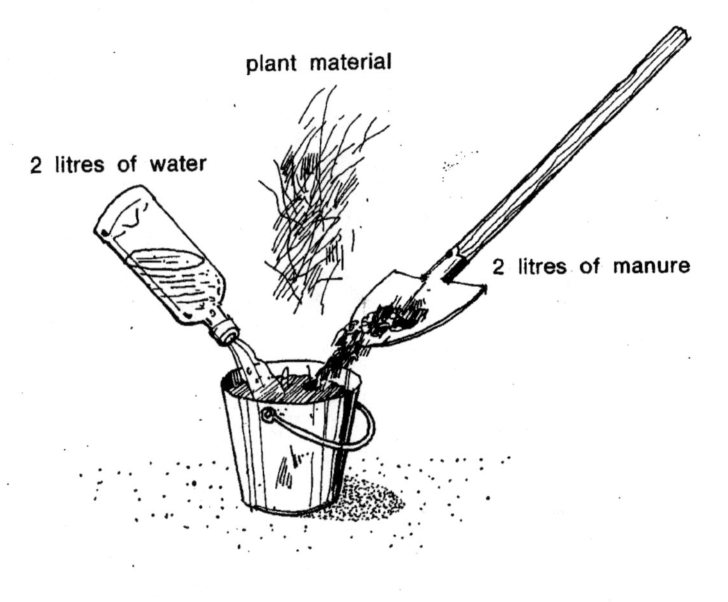

68. We call this mixture a **starter.**
A starter helps the biogas unit
to make gas sooner.

69. Pour the starter mixture
into a container
which holds a little more than 4 litres.
You can use a bottle or a jug
but do not close it,
leave it open.

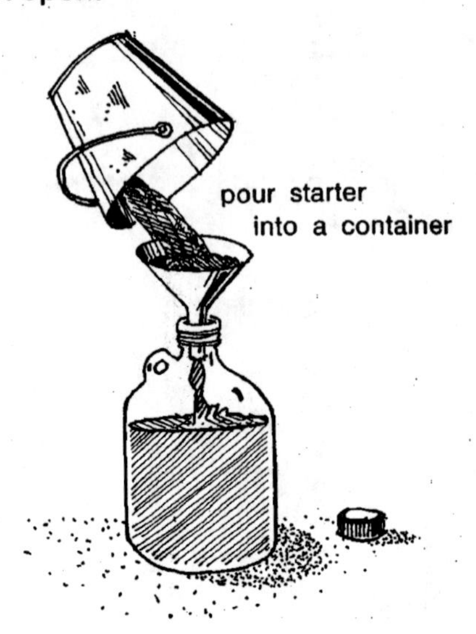

70. Keep the starter warm
and shake the container
three or four times each week
to mix the contents.
In about two months
it will be ready to use.

Putting waste into your biogas unit

71. Now you are ready to put the waste into your biogas unit.
    Put the large drum open end up where you want the unit to be.
    Put the small drum next to it with the gas outlet up.

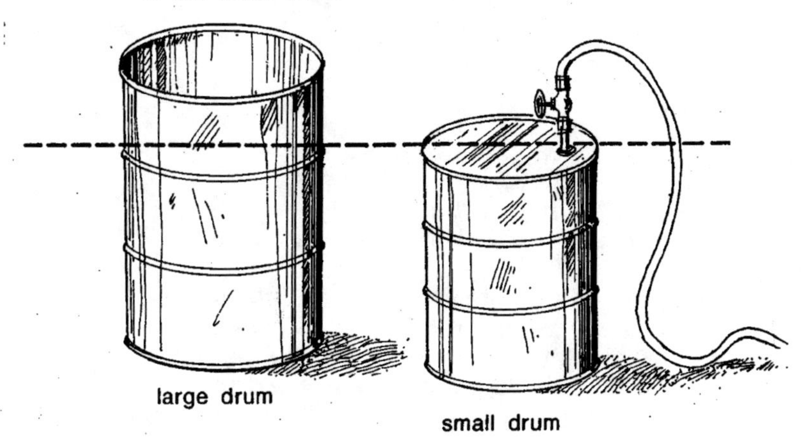

72. Now put the waste and water you are going to use into the large drum.
    Put 3 buckets of waste and 3 buckets of water into the large drum and stir it well.

73. Now put another 3 buckets
of waste and 3 buckets of water
into the large drum
and stir all of the waste mixture again.

74. Put more waste and water into the large drum,
stirring well each time,
until the waste mixture in the large drum
is level with the top of the small drum.
The drawing on page 29 will show you how.

75. Stir the starter
you have made,
(see Items 67 to 70)
into the waste mixture
in the large drum.
The starter
which has already
begun to work
will help you
to make gas sooner.

76. Now open the valve or clamp
or untie the gas line
of the small drum
to let out the air.
Push the small drum down
into the waste mixture
until it touches the bottom
of the large drum.

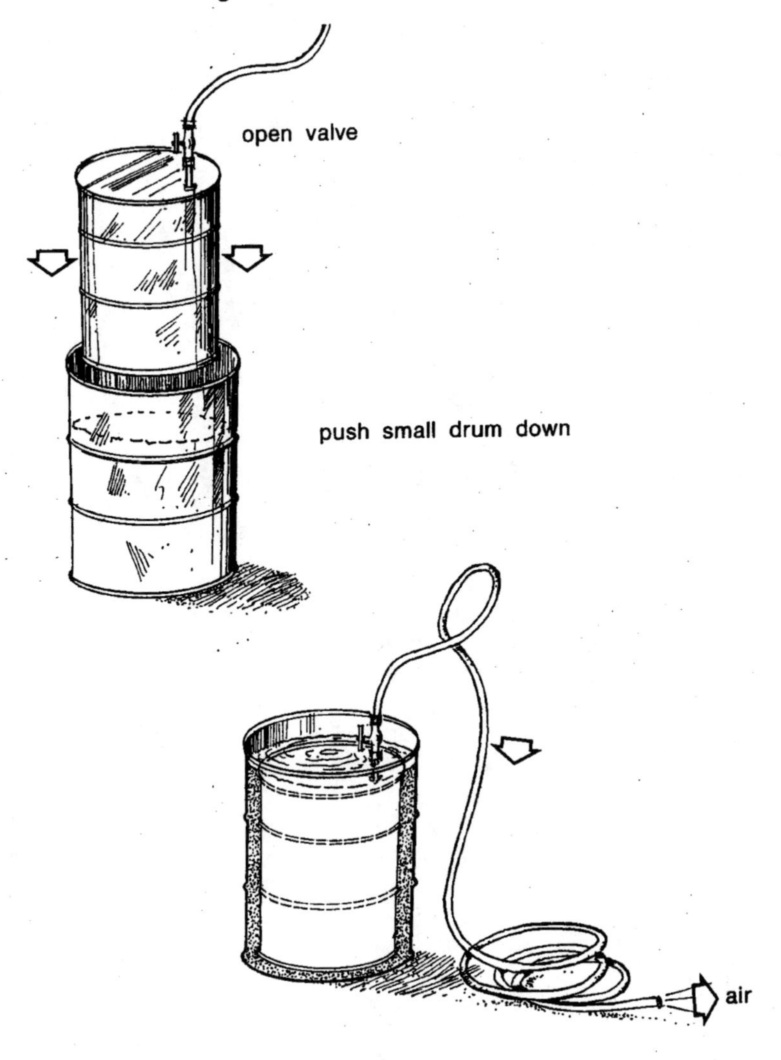

77. The small drum
must be full of waste mixture.
It must be full to the top
so that there will be no air in it.

78. You can be sure that the small drum
is full to the top
if you can see that the waste mixture
inside the large drum
rises a little above the top edge
of the small drum
when it has been pushed down.

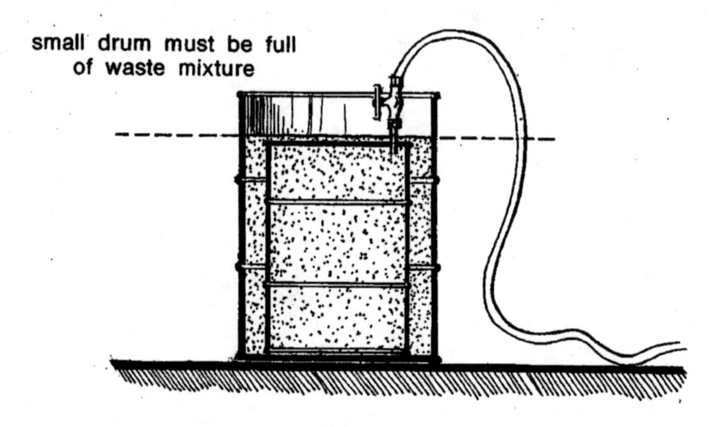

79. If it does not rise above the top edge
of the small drum,
take the small drum out
and put a little more waste and water
into the large drum.
Then put the small drum back
and push it down into the waste again.

80. When you are sure
that the small drum
is full of waste mixture to the top,
close the valve
or clamp or tie the gas line
so that you will keep out air
and begin to collect gas.

81. You can tell that the waste mixture
in your simple biogas unit
has begun to rot and make gas
when the small drum begins to rise.
This means that gas is being collected.

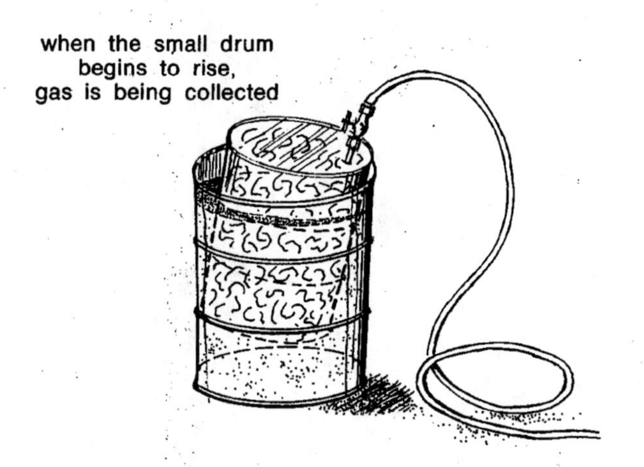

82. If you find that gas is leaking
from the small drum
after the biogas unit has begun to work,
seal the leaks with tar, mastic or paint.
If the gas is leaking
around the gas outlet or valve,
tighten the outlet or valve again
and coat the joints
with tar, mastic or paint.

83. A good way to check for leaks after the biogas unit has begun to work is to put soapy water on the small drum and on the joints of all the parts and lines. If you see bubbles anywhere, you will know that there is a leak. Seal the leaks as you were told in items 53 and 54.

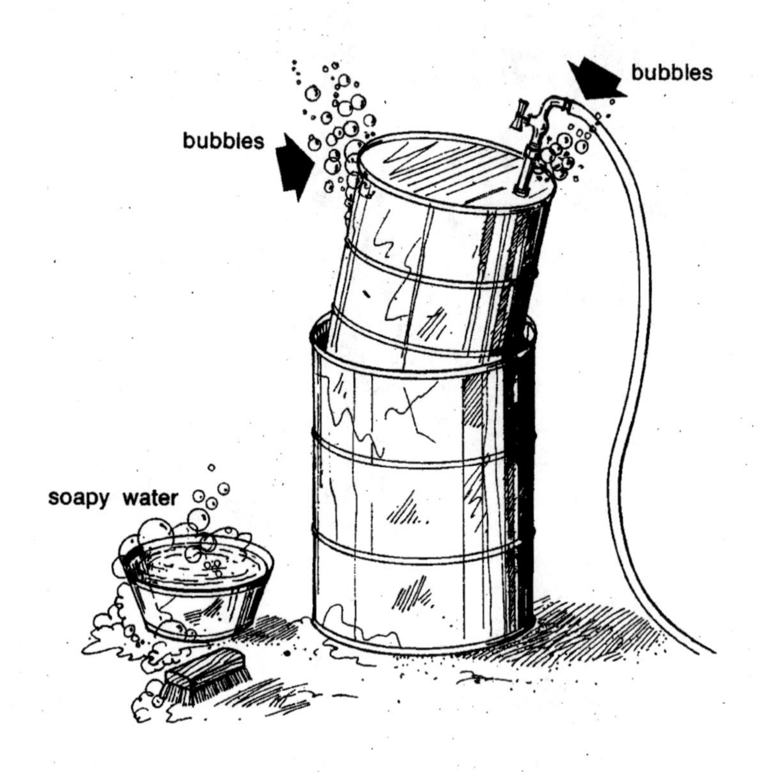

TIME

84. It may take up to three weeks
    or even a month
    for the waste in your biogas unit
    to start making gas.
    After that, gas will be made
    for about eight weeks.

85. During these eight weeks,
    half of the gas will be made
    in the first two or three weeks
    and the rest
    in the last five or six weeks.

86. If you find that not much gas
    is being made in the last weeks,
    empty the unit and start again.

87. If the temperature where you are often falls below 15°C, you will have to keep the waste mixture in your biogas unit warm.

88. If you put your biogas unit under the ground or partly under the ground, this will help to keep the waste warm.

put the large drum partly under the ground

89. You can keep the waste mixture in your biogas unit warm by putting leaves, grass, straw or maize stalks around the large drum.

90. You can also keep it warm
by adding a bucket or two
of poultry droppings
mixed with other waste
to the waste already in the large drum.
Use one part of poultry droppings
to three parts of other waste.

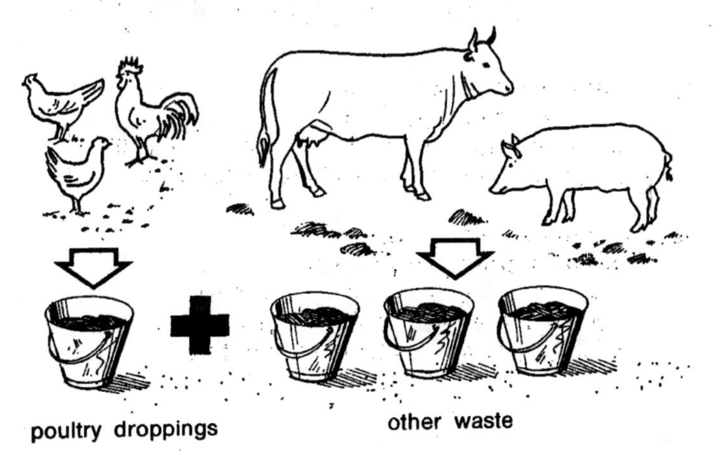

STIRRING THE WASTE MIXTURE

91. Sometimes a layer of scum
may form on top of the waste mixture
in your biogas unit.
If this happens, less gas will be made
and the small drum may not rise.

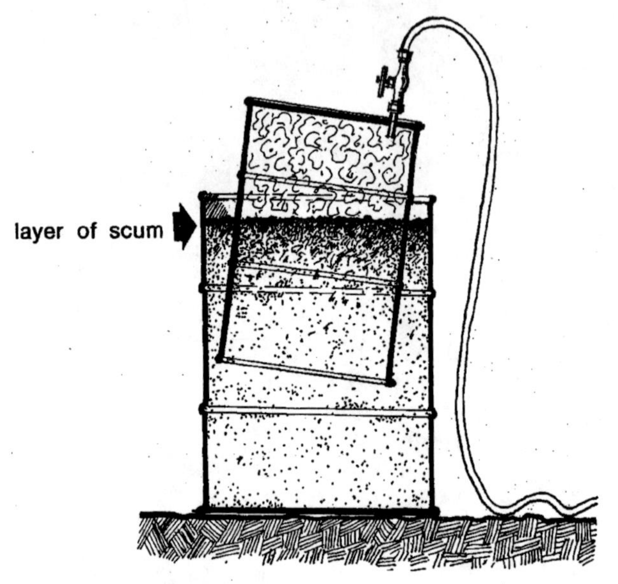

92. If the waste is well mixed
before it is put into the unit,
there will be less chance for scum to form
and your biogas unit will make gas well.

93. If you use plant materials,
scum is more likely to form
than if you use only animal manure.
You will need to stir the waste mixture
in your biogas unit from time to time.

94. You can break up a scum layer
by stirring or shaking the waste mixture
after it is in the unit.
You must do this
without opening the small drum
and letting out the gas
or letting in the air.

95. During warm weather,
the waste mixture in your biogas unit
may become too thick
and little gas will be made.

96. If this happens,
add a bucket of water to the unit
and stir the waste to thin it.
If after a few days
no gas is being made
and the waste is still too thick,
add another half-bucket of water
and try again.

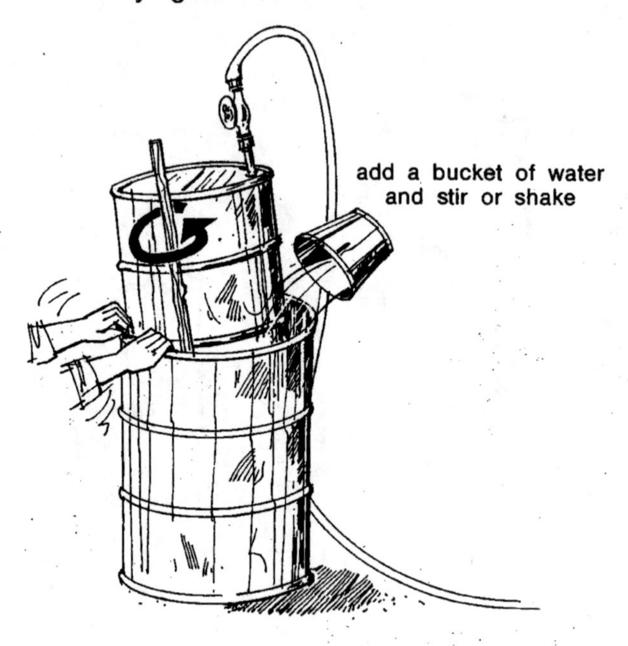

97. If a lot of very hard scum
forms on top of the waste mixture
and no gas is being made,
take out all the waste mixture,
clean the unit and start again.
Do not throw the waste mixture away,
use it for fertilizer.

WHEN THE GAS IS MADE

98. **Do not burn**
    the first gas that is made.
    It may have air in it
    and could explode.

99. A few days after the small drum
    has begun to rise,
    open the valve or clamp
    or untie the gas line
    and let out all of the gas
    that has been collected.

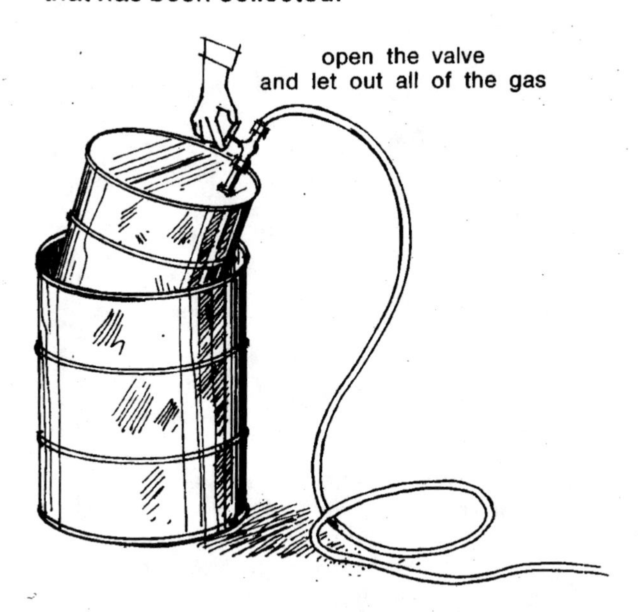

100. While you are letting the gas out,
     be very careful not to have fire
     near the biogas unit.

101. To let the gas out,
push the small drum
back down into the waste mixture
in the large drum.
This will force all gas and air
out of your biogas unit.
Then, close the valve
or clamp or tie the gas line
and your biogas unit
will begin to collect gas again.

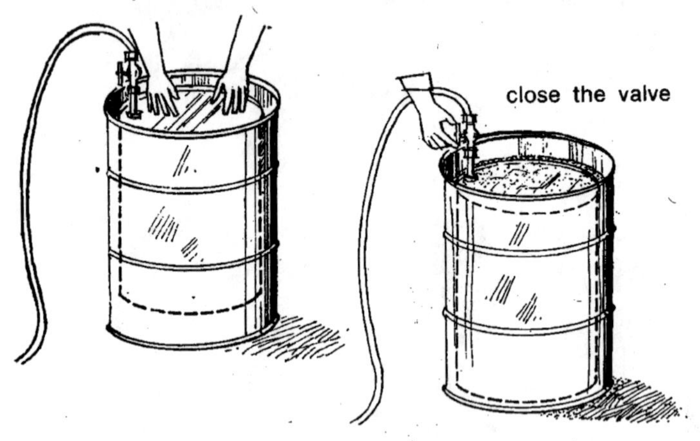

102. If you have done this carefully,
the next gas that is made
will have no air in it
and it will be safe to burn.
You can burn all the biogas
that is made after this.
Do not open the unit again
until all the gas has been made.

103. After all the gas has been made, take the unit apart and empty out the fertilizer. Keep about 4 litres of the fertilizer to be used as a starter for the next time.

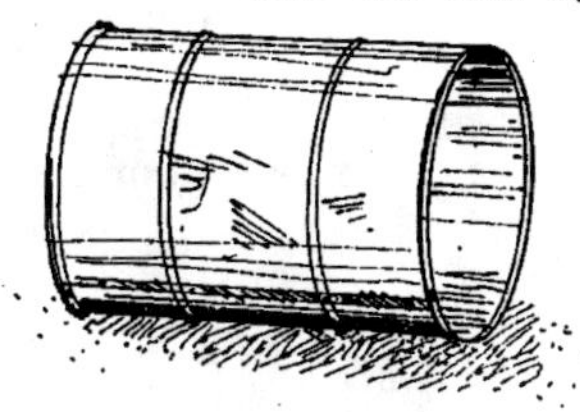

104. Clean the unit and check for leaks.

105. Now fill the unit
with new waste material
and add the starter.
Close the unit tightly
and it will begin to make gas again.

fill with new waste material

close the unit tightly
and push the small drum down

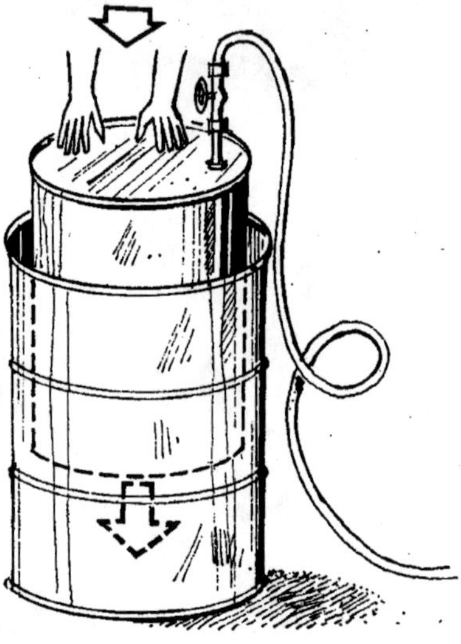

106. **Remember,**
every time you start again,
do not burn
the first gas that is made.

107. The best way to use the biogas
that you make
with your small biogas unit
is for cooking.
When your unit is working well,
it will make enough gas every day
to cook your evening meal.

108. You can use biogas
with almost any ordinary gas-burner,
if you adjust the burner
so that the right amount of air
is mixed with the biogas.

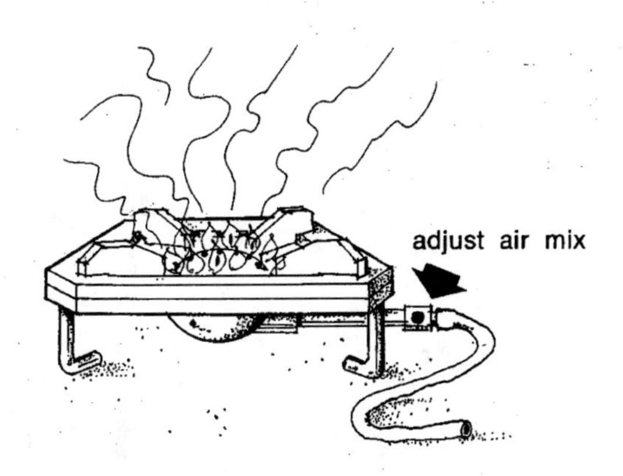

109. If there is too much air,
the flame will be blown out
and the biogas will not burn.
If there is not enough air,
the flame will be yellow,
the biogas will not burn well
and will not give enough heat.

110. When there is the right amount of air
and the biogas is burning well,
it will burn with a blue flame.
By letting more or less air
into the burner,
try to make the flame
as blue as you can.

111. Sometimes the flame
may begin to turn yellow
after it has been burning well.
This may mean that the burner
has become full
of a black material called soot.

112. If this happens,
clean the burner very carefully
and clean all the holes in the burner
with soap and water.
Dry the burner well.
This may help your biogas
to burn well again.

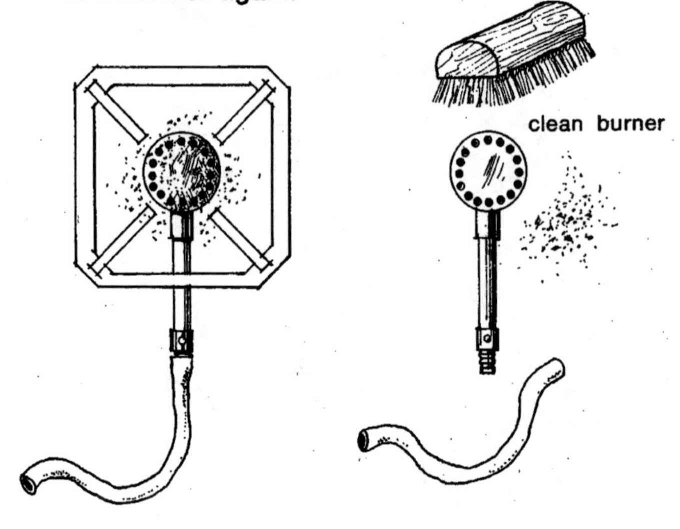

113. If the flame is not steady,
or if it is weak
when there is still gas in the unit,
this may be because there is water
in the burner or the gas line.

114. Shut off the gas
at the small drum
and take off the burner.
Empty out any water
that is in the gas line
or in the burner.
Then put the burner back,
turn on the gas
and light the burner again.

USING THE FERTILIZER

115. You have already learned
that when all the gas has been made,
the material that is left
in your biogas unit
is a very rich fertilizer.

116. It does not have a bad smell,
and the parasites
that were in the animal wastes
and the weed seeds
that were in the plant wastes
are no longer harmful.

117. You can spread this new fertilizer
on your fields
to help your plants grow well.

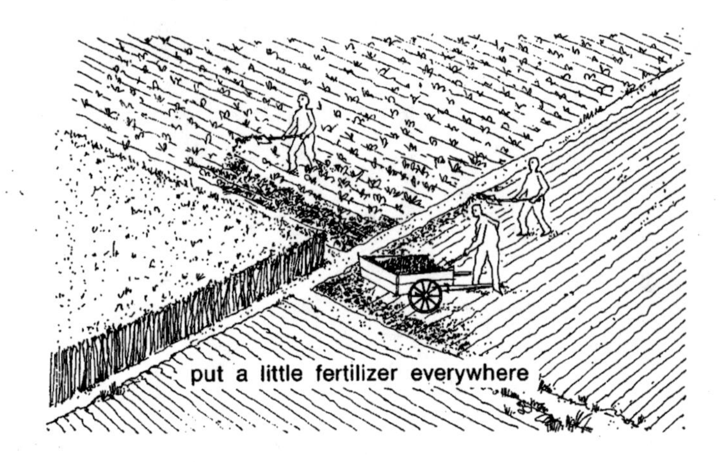

118. Do not put the fertilizer
all in one place.
Put a little of it everywhere
on your fields.
In this way,
all of your plants
will grow better.

TAKING CARE OF YOUR BIOGAS UNIT

119. **Always be very careful**
when you are near a biogas unit
because gas may be leaking.

120. **Never build a fire near the unit,**
smoke, or even light a match
near the unit,
because if gas is leaking
it may explode.

121. **If biogas is leaking**
and you breathe in too much of it,
it can make you very sick.

122. Check your biogas unit
and gas lines often
to be sure that there are no leaks.
Items 49 to 57
and items 82 and 83 have told you
how to find and stop leaks.

123. After some time,
rust will start to appear
on the inside of your unit.

124. Once a year
you should take the unit apart
and clean and paint
the metal gas holder
and all other metal parts.

125. You can use paint
which is used to protect metal
or coat the metal parts with tar.

MAKING MORE BIOGAS

126. After you have made biogas a few times
with your small biogas unit,
and have used it for cooking,
you may find
that you could use more gas
if you had it.

127. The easiest way
for you to make more biogas
is to build one or more biogas units
just like your first one.

128. If you can get more oil drums,
pipe, valves and gas lines,
and if you have enough waste materials,
you can build and run
several small biogas units
and get gas from all of them.

129. When you have several biogas units,
you can connect them to your gas line
by using T-pieces.
The drawings show you a T-piece
and how to connect several units
to the same gas line.

T-piece

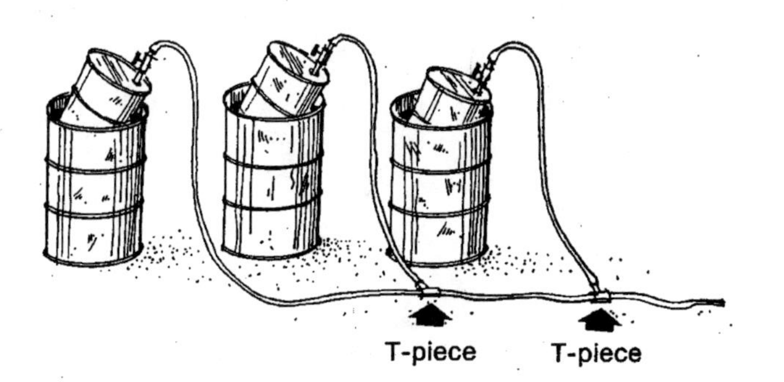

130. When you have several biogas units
fill them with waste at different times
so that when all of the gas in one unit
has been made,
you will still get gas
from another unit which is working.

131. In this booklet you have learned
how to build a small biogas unit
and how to make your own biogas.
You have also learned
that you can make more biogas
by building several small biogas units.

132. But there are still other ways
to make more biogas.
You can build an improved small unit
or you can build a different kind of unit
which is bigger and better
and will give more biogas.

133. You will learn about these ways
in a later booklet.

better farming series 32

biogas 2

building a better biogas unit

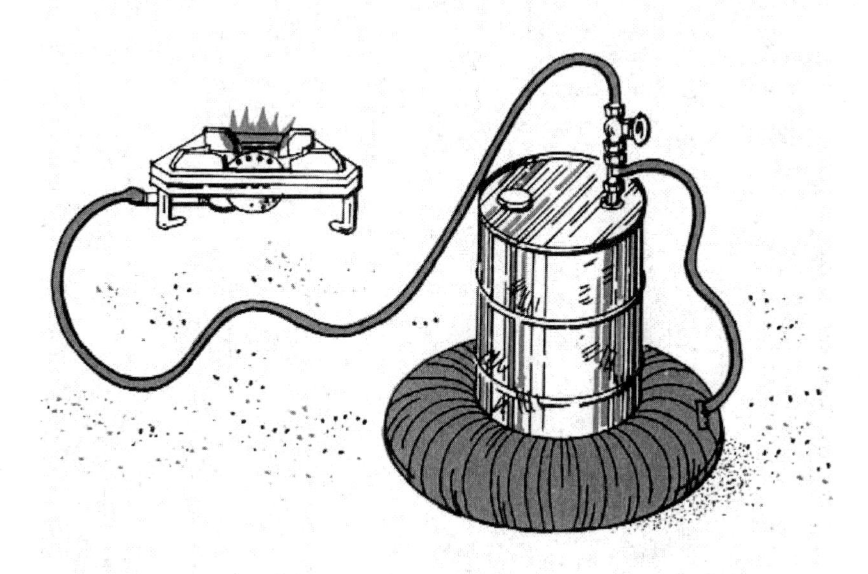

FOOD AND AGRICULTURE ORGANIZATION OF THE UNITED NATIONS

better farming series 32

biogas 2

building a better biogas unit

FOOD AND AGRICULTURE ORGANIZATION OF THE UNITED NATIONS
Rome 1986

FAO Economic and Social Development Series No. 3/32

PREFACE

The first twenty-six volumes in FAO's Better Farming Series were based on the **Cours d'apprentissage agricole** prepared in Côte d'Ivoire by the **Institut africain de développement économique et social** for use by extension workers. Later volumes, beginning with No. 27, have been prepared by FAO for use in agricultural development at the farm and family level. The approach has deliberately been a general one, the intention being to constitute basic prototype outlines to be modified or expanded in each area according to local conditions of agriculture.

Many of the booklets deal with specific crops and techniques, while others are intended to give the farmer more general information which can help him to understand **why** he does what he does, so that he will be able to do it better. This booklet was added to the series owing to the favourable comments received on Booklet No. 31, **Biogas: what it is; how it is made; how to use it.** Both booklets have been based on published works by researchers and experimenters in small-scale biogas production in Africa, Asia, Europe and North America.

Adaptations of the series, or of individual volumes in it, have been published in Amharic, Arabic, Bengali, Creole, Hindi, Igala, Indonesian, Kiswahili, Malagasy, SiSwati, Thai and Turkish.

Requests for permission to issue this manual in other languages and to adapt it according to local climatic and ecological conditions are welcomed. They should be addressed to the Director, Publications Division, Food and Agriculture Organization of the United Nations, Via delle Terme di Caracalla, 00100 Rome, Italy.

OUTLINE OF THE BOOKLET

INTRODUCTION

1. You have already built
   one or more biogas units
   like the one described
   in the Better Farming Series
   Booklet No. 31; **Biogas: what it is; how it is made; how to use it**.

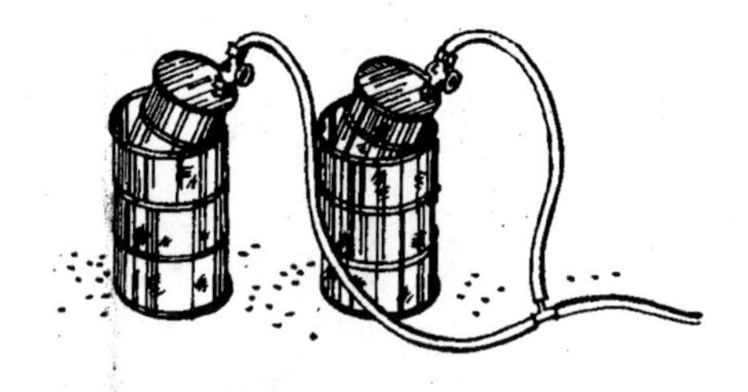

2. When you first began
   you found that you had to learn
   a lot of new things
   in order to make your unit work.

3. However, little by little
   you have learned more and more
   through your own experience.

4. When your first gas was made
   you used it for cooking.
   You found that cooking with gas
   was cleaner, easier and faster
   than cooking with kerosene,
   charcoal or fuelwood.

5. Now that you know more about biogas and how it is made, let us look at another way to make biogas even better.

6. In this booklet you will learn how to build and use a **better** small biogas unit like the one shown below.

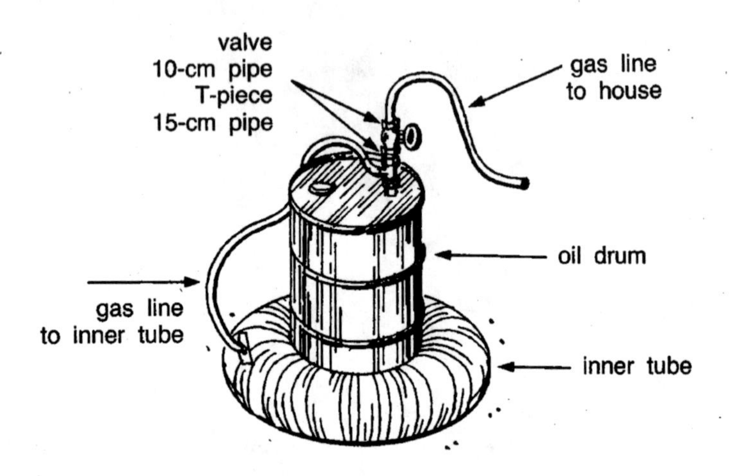

7. It too has an oil drum for a waste holder and, like your first small unit, **all** the waste is put in **at one** time when you begin.

8. However, the new unit is closed. A closed unit is cleaner. You cannot smell the waste after you have put it in as you could with your old unit.

9. In addition, with your old unit some of the gas was lost from around the open sides of the oil-drum waste holder.

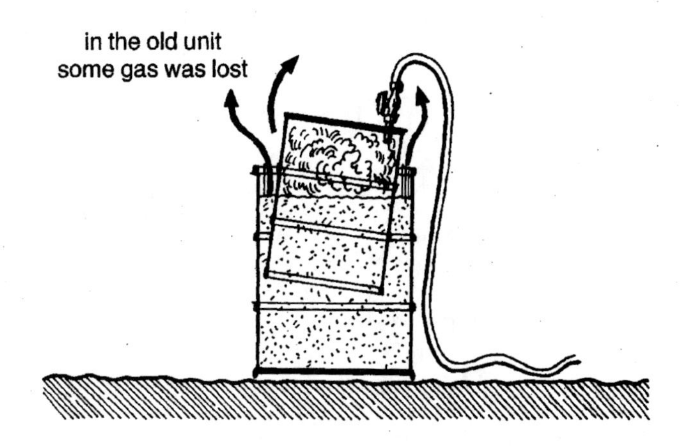

10. Since the new unit is closed
you will **not** lose any gas.
You can collect it all
so you will have more gas to use.

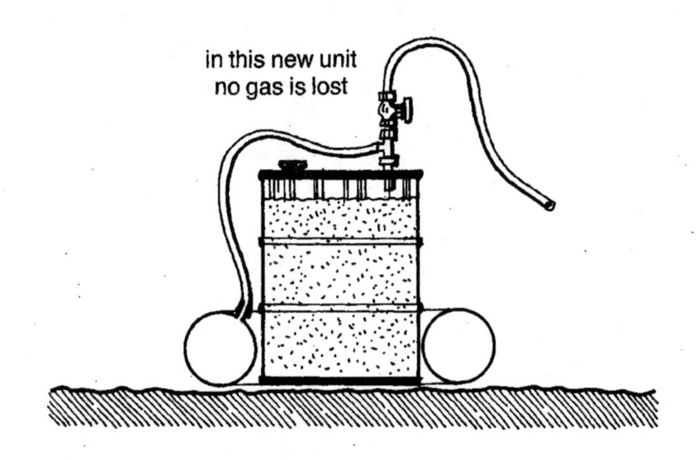

11. Notice in the closed unit
(see the drawing above)
that the oil drum is filled
nearly to the top with waste.
There is little space to hold gas.

12. So, you will need something
    to collect the gas.
    In the new unit,
    the gas holder is a used inner tube
    as you can see in the drawing
    on page 2 in this booklet.

13. This new biogas unit
    looks much like your old unit
    and it works in much the same way.
    You already know a lot of the things
    that you need to know
    to put this new unit together.

14. However, before you begin
    it would be a good idea
    to read Booklet No. 31 again.

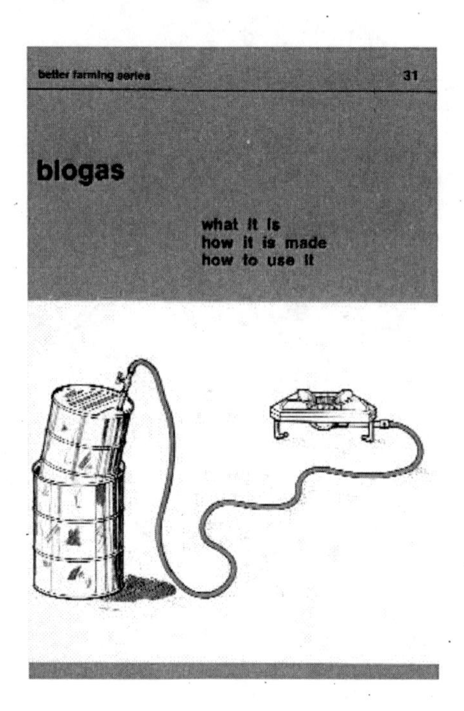

HOW TO BUILD
A BETTER SMALL UNIT

You will need

- an oil drum of about 200 litres,
  to hold the waste

200-litre drum

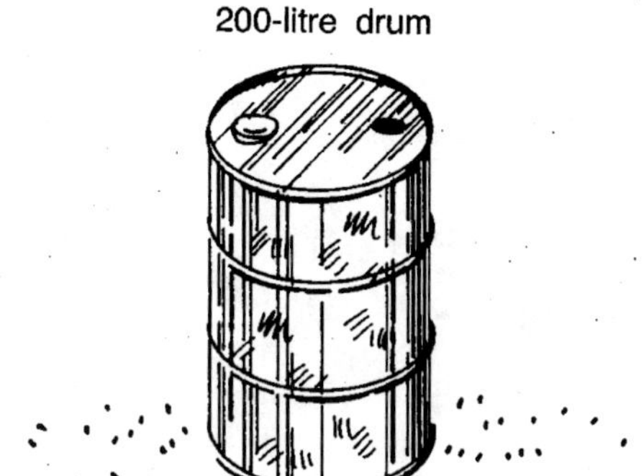

- a piece of pipe
  about 15 centimetres long
  and about 2 centimetres in diameter
  to fit the oil drum, for the gas outlet

gas outlet

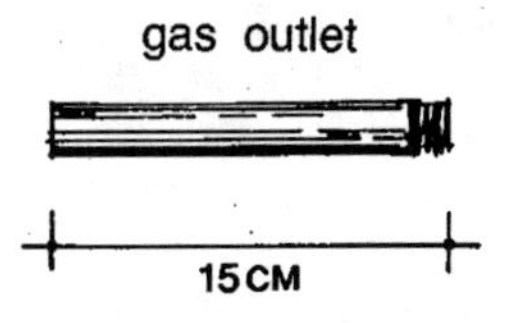

- a pipe T-piece,
  to connect the gas outlet
  to the inner tube

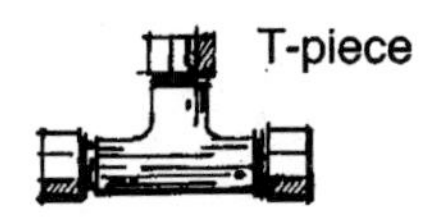

T-piece

- a piece of pipe
  about 10 centimetres long
  to fit the T-piece

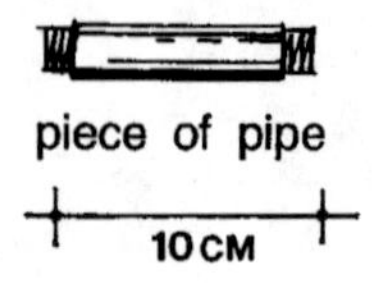

- a valve to fit
  the 10-centimetre pipe

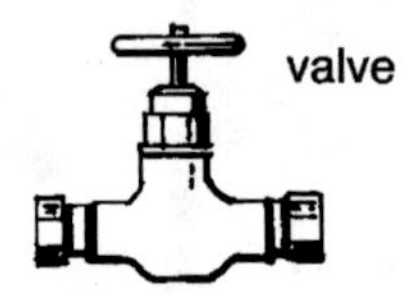

Note

The drawing on page 14 in this booklet shows you how these pipe fittings are attached to the oil drum.

- at least 12 metres
  of rubber or plastic tube,
  about 2 centimetres in diameter,
  for the gas lines

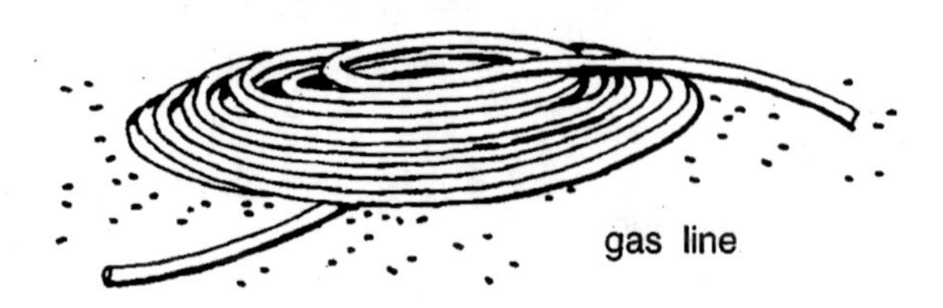

- one or more inner tubes,
  to collect the gas

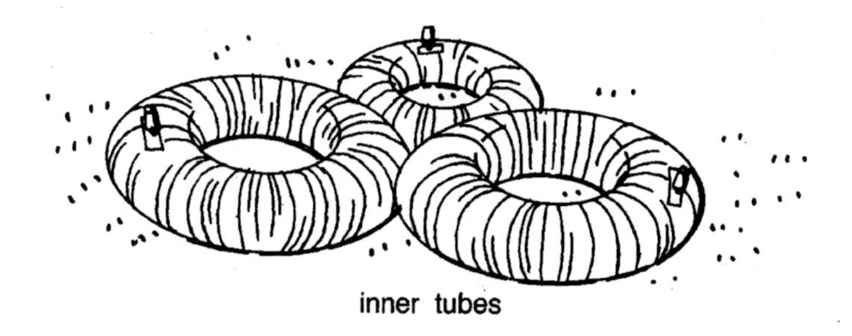

inner tubes

- if you are using
  more than one inner tube,
  you will also need
  one or more small T-pieces
  to connect the inner tubes.

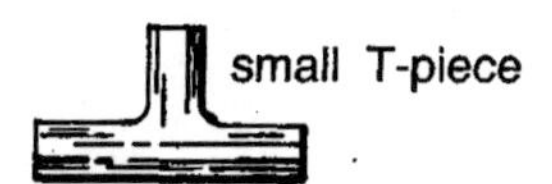

small T-piece

15. The oil drum should have
    one hole for putting in the waste
    and another hole for the gas outlet.
    Many drums have threaded holes
    with threaded plugs to close them.

16. Try to find an oil drum
    with threaded holes in the top.
    That way it will be easier
    to build this unit
    and to make it airtight.

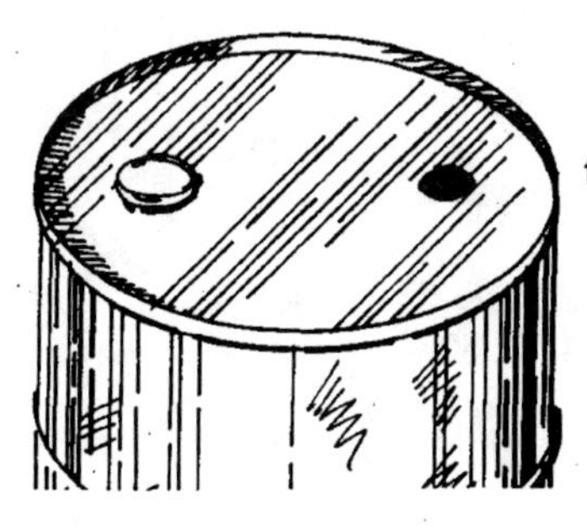

drum with two threaded holes

Cleaning the oil drum

17. Begin by cleaning the drum
    inside and outside
    to remove all oil and grease.

18. Take the metal plugs
    out of the holes
    and put them carefully aside,
    so that you can find them later.

19. First clean the **Inside** of the drum.
    Pour in a bucket or two of warm,
    soapy water or other cleaner.
    Then close all of the holes.

20. Put the drum on its side.
    Roll it back and forth
    so that the soapy water or cleaner
    can wash the whole inside.
    Then open the holes
    and empty out the water.

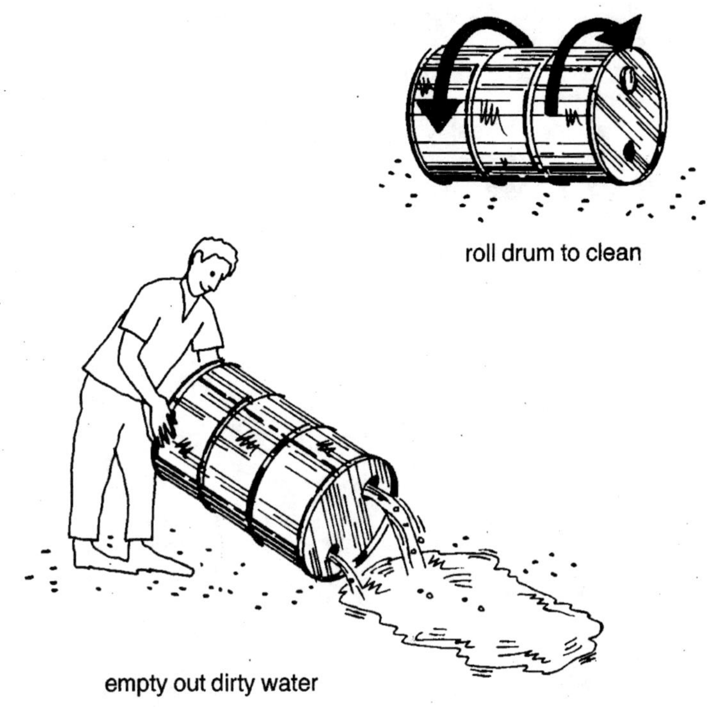

roll drum to clean

empty out dirty water

21. Continue to wash the inside
    of the oil drum
    with soapy water or cleaner
    until it is completely clean.

22. You can tell the oil drum
    is clean when the water
    you empty out is clean.

23. When you are sure
    that the inside is clean,
    pour in three buckets of fresh water
    and roll the drum
    back and forth once more.
    This is to rinse out any soap
    or cleaner that is still inside.
    Then empty it out again.

24. Now clean the **outside** of the drum with a brush and soapy water or cleaner. Rinse it with fresh water.

25. Open the holes in the top and put the drum on stones with the top down. Let it drain and dry.

clean the outside

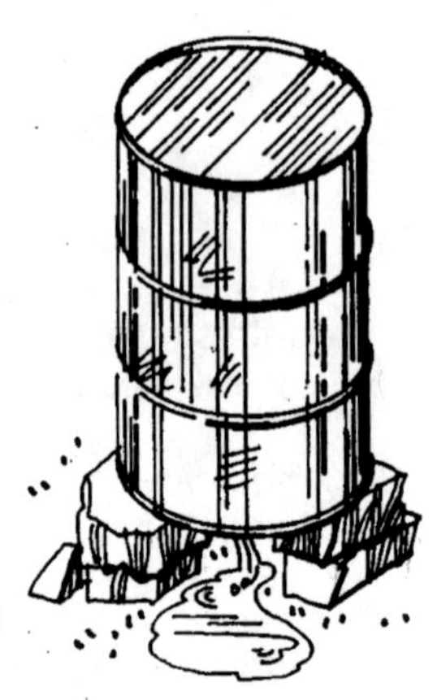

raise the drum on stones and let it drain and dry

26. When the drum is dry inside and out you are ready to begin.

Where to put your biogas unit

27. Before you build
    your new biogas unit
    you should decide where to put it.
    Items 26 to 34 in Booklet No. 31
    will tell you where.

28. However, **do not**
    put this unit underground.
    If the unit is underground
    you will not be able to shake it
    to break up the scum
    (see Items 109 to 113 in this booklet).

Preparing the oil drum

29. If your oil drum
    has a hole in its side,
    close it tightly.
    You can use a threaded metal plug
    or weld a piece of metal
    over the hole.

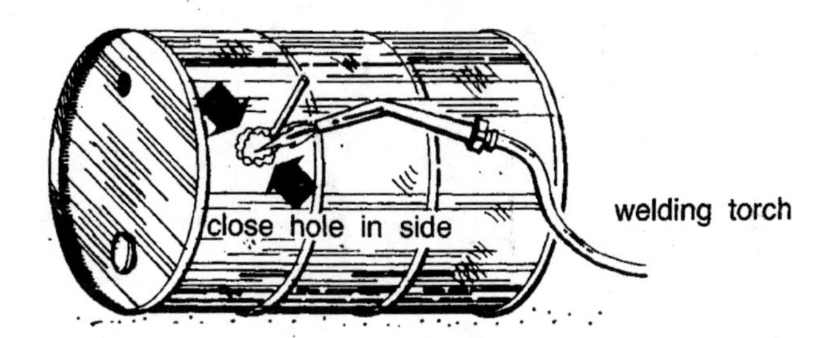

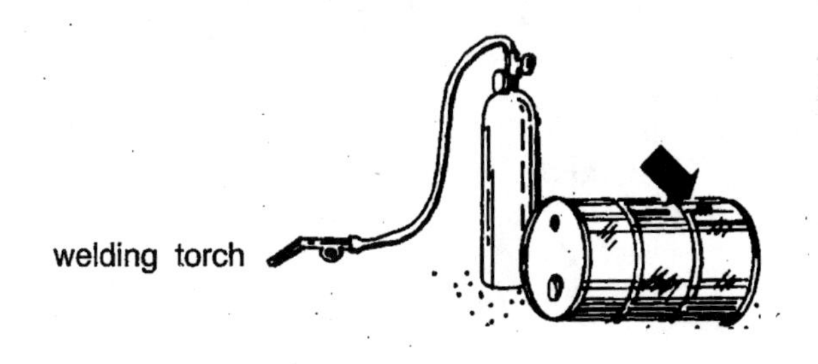

30. Now you are ready
    to put the gas outlet
    in the top of the drum.

31. If your drum
    has two holes in the top,
    use the smallest one
    for the gas outlet.
    Save the largest one
    for putting in the waste.

32. The gas outlet
    is made from a piece of pipe
    about 15 centimetres long
    and about 2 centimetres in diameter.
    However, it should fit the hole
    in the drum.

33. If the hole **is** threaded,
    use an outlet pipe
    that is threaded on both ends.
    Screw it tightly into the hole.

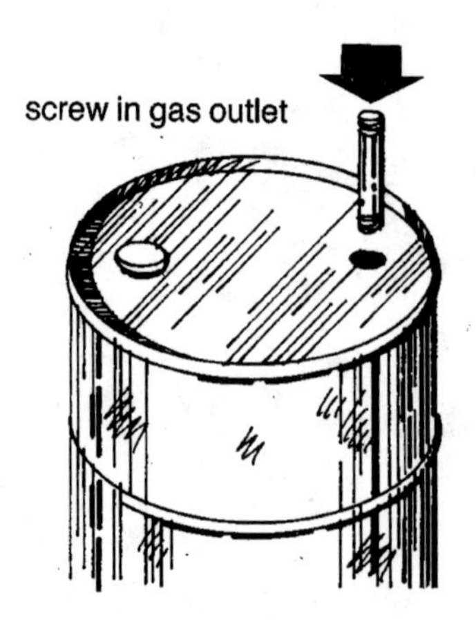

34. If the hole **is not** threaded,
    use an outlet pipe
    that is threaded on one end.
    Weld it into the hole
    with the threaded end up.

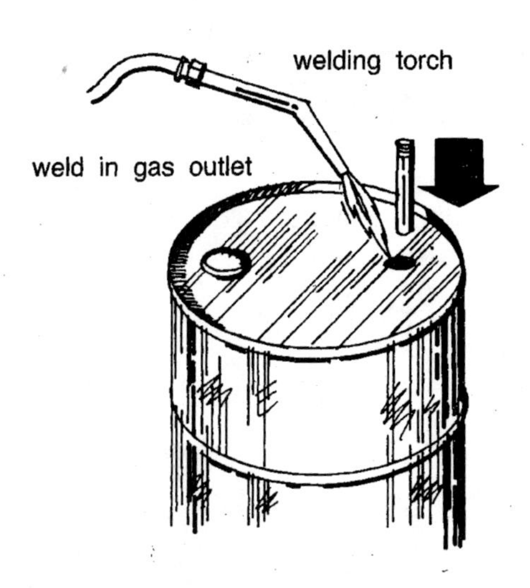

35. If there is **only one** threaded hole in the top of the oil drum, use it to put in the waste.

36. Then you will have to cut a hole about 2 centimetres in diameter for the gas outlet. Weld in a pipe that is threaded on one end, as shown in the drawing at the top of this page.

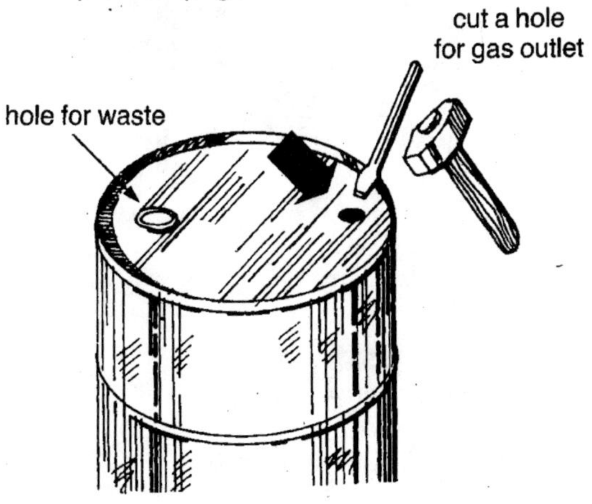

37. Now you are ready
    to attach the pipe T-piece,
    the 10-centimetre piece of pipe
    and the valve.

38. The valve you use must be airtight
    so that it will not leak gas.
    You must be sure
    to screw all of these pieces
    tightly to the gas outlet.

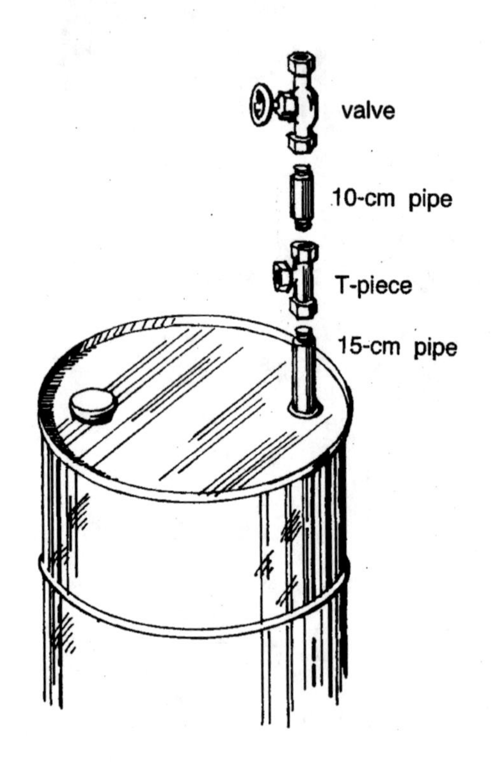

Note

If you do not have a valve,
you can tie or clamp the gas line
to stop the flow of gas
(see Item 48 in Booklet No. 31).

Testing for leaks

39. Now you are ready
    to test the drum for leaks.
    To make biogas,
    the drum must be airtight.

40. To test for leaks,
    open the valve,
    take out the metal plug
    in the waste hole
    and fill the drum with water.

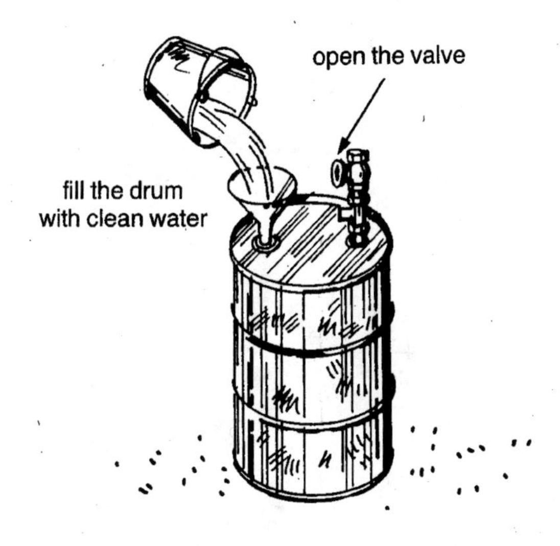

41. **Be sure** to fill it to the top.
    Then close the valve
    and put the metal plug back
    in the waste hole.

42. Use a piece of cloth
    to dry any water
    that you have spilled
    on the outside of the drum.

43. If you see water leaking
from anywhere on the drum,
mark the place of each leak.

44. Then turn the drum over
on its side.
When the drum is full of water
it is very heavy,
so ask someone to help you.

45. Now check for leaks
on the top part of the drum
and around the gas outlet,
T-piece and valve.
If there are leaks here,
mark them too.

46. Then open the valve,
take out the metal plug
and empty out the water.
Raise the drum on stones
with the top down
so it can drain dry.

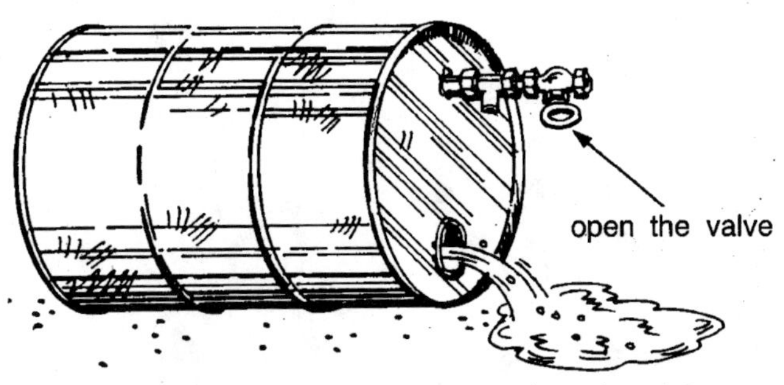

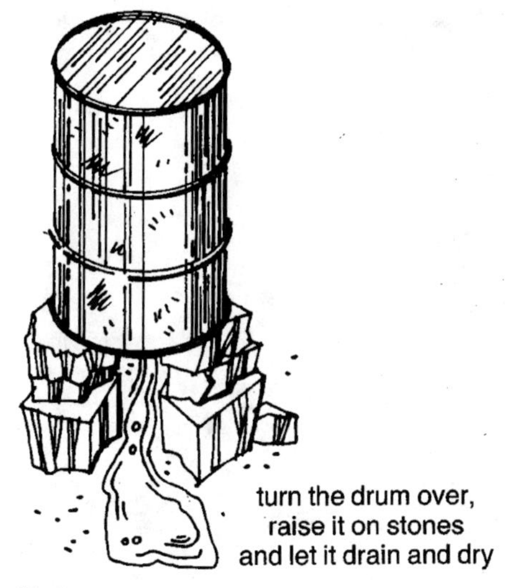

Note

When you put the drum on stones,
be very careful
not to damage the gas outlet,
the pipe T-piece or the valve.

47. Seal the leaks by coating them with tar, mastic or paint. If there are any leaks around the gas outlet, tighten the outlet, T-piece and valve again. Coat the joints with tar, mastic or paint.

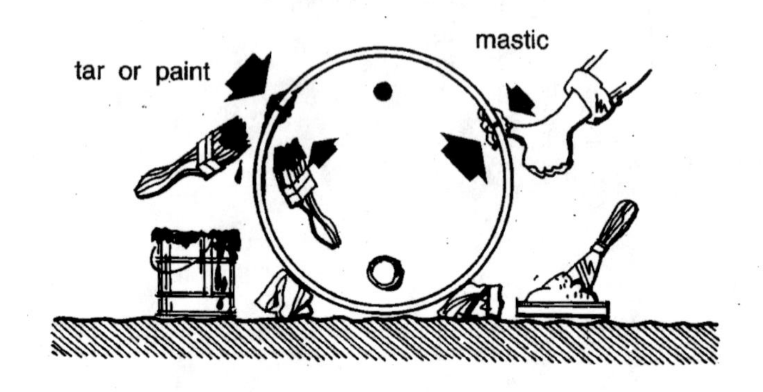

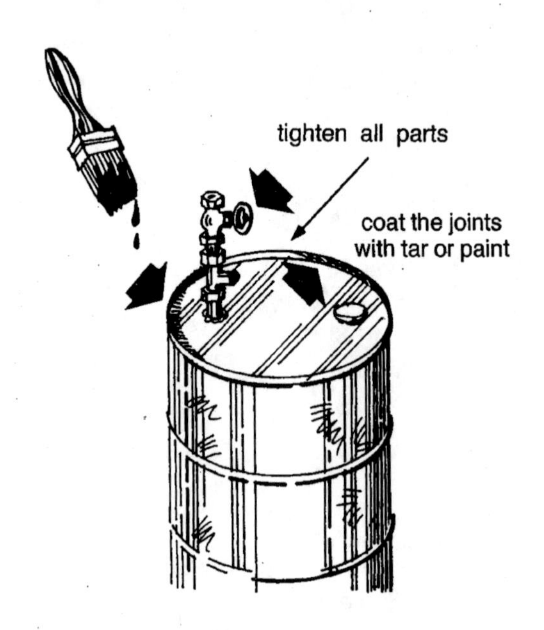

48. When the sealing is dry,
fill the drum with water again.
Check that all the leaks are sealed.
If the drum still leaks,
empty out the water
and let it dry.

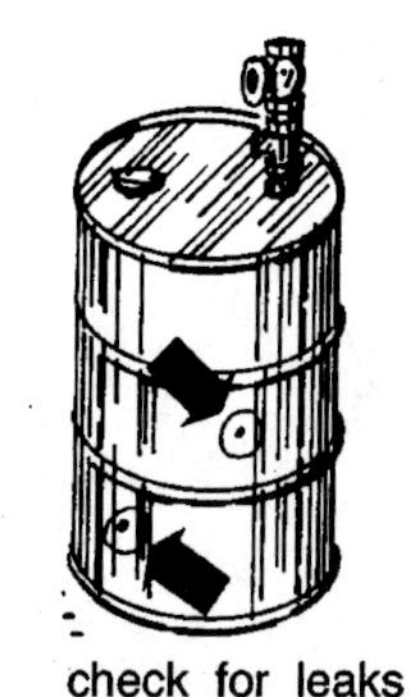

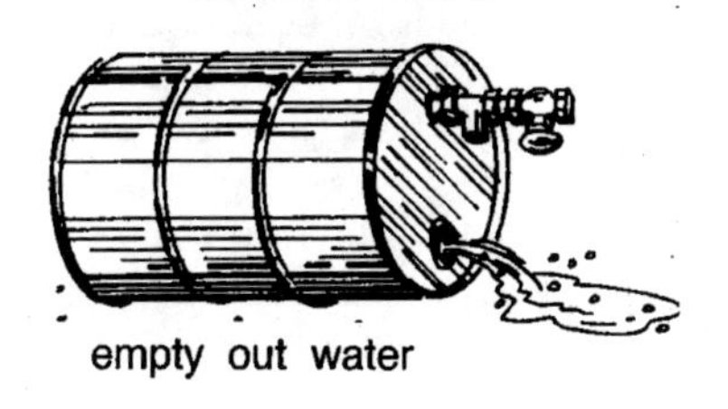

49. Put some tar or paint
inside the drum.
Then turn the drum
around and around
to coat the inside of it.

50. Fill the drum with water again.
If it still leaks,
start all over again.
**It is very important
to seal all leaks carefully.**

51. When the drum is well sealed
and no longer leaks,
let it dry completely.
Now you can begin to prepare
the gas holder.

Preparing the gas holder

52. As you have been told,
this unit has an inner tube
which holds the gas.

53. If you can get
a large truck or tractor inner tube,
you will need only one.

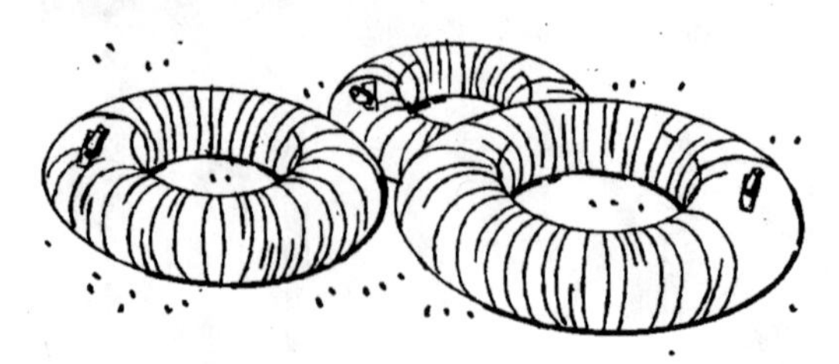

54. You can also use
tubes from automobiles.
However, these are smaller
so you will need two or even three.

55. Try to get a large tube
because it is easier to attach
one large tube
than two or three small ones.

56. First, check each tube
that you are going to use
for leaks.
Each tube must be airtight.
To check a tube for leaks
fill it with air.

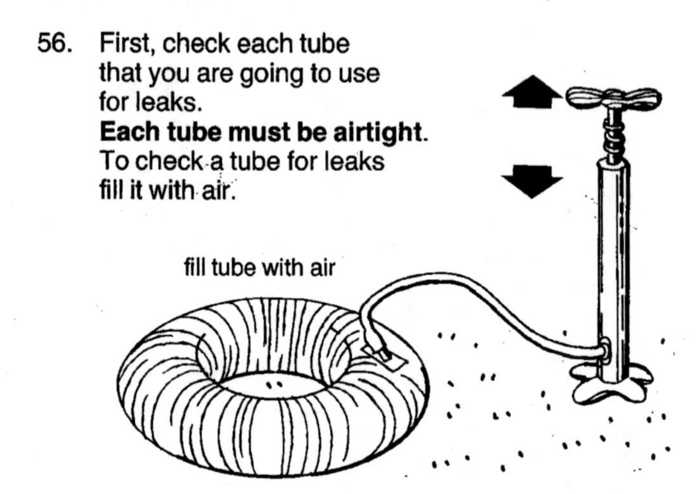

57. Then put the tube in water.
You can put it in a pond
or in a quiet stream.

58. Turn the tube slowly under the water.
Look for air bubbles.
If you see any bubbles,
mark each place on the tube.

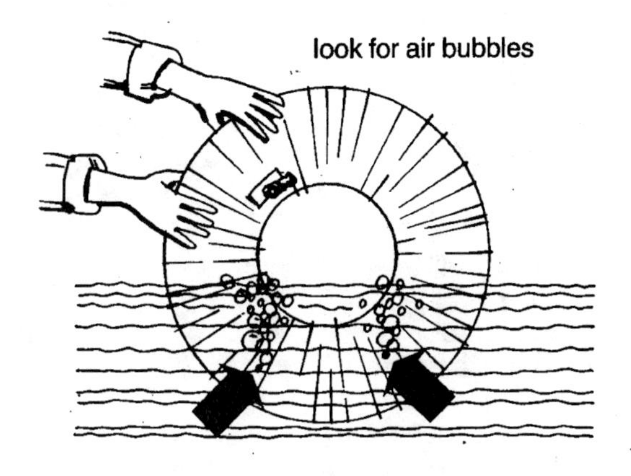

59. Let the tube dry.
When it is completely dry,
repair all of the leaks.

60. Fill the tube with air
and put it in the water again
to make sure that you have
repaired all of the leaks well.

put in water again
to check for leaks

repair all leaks

61. If there are still leaks,
start all over again.
**It is very important
to seal all leaks carefully.**

62. When **all** the leaks are sealed
let **all** of the air out.
To do this, unscrew the cap
of the air inlet and remove the valve.

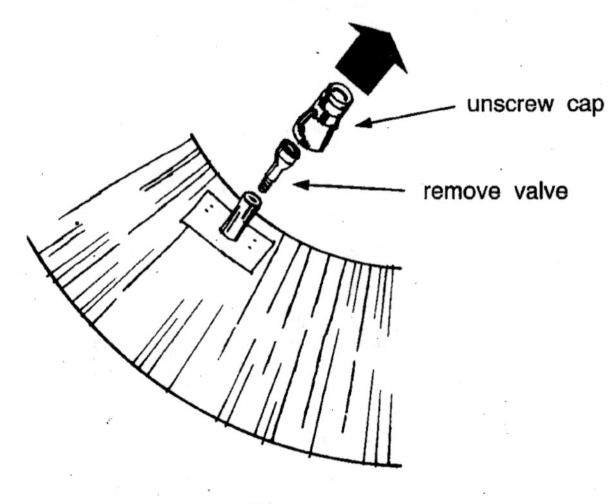

63. Roll the tube very tightly.
If you have a smooth, round pole,
you can roll the tube around this.

64. When the tube is tightly rolled
and there is **no** air in it,
screw the cap on the air inlet.
This is to keep more air
from getting inside.

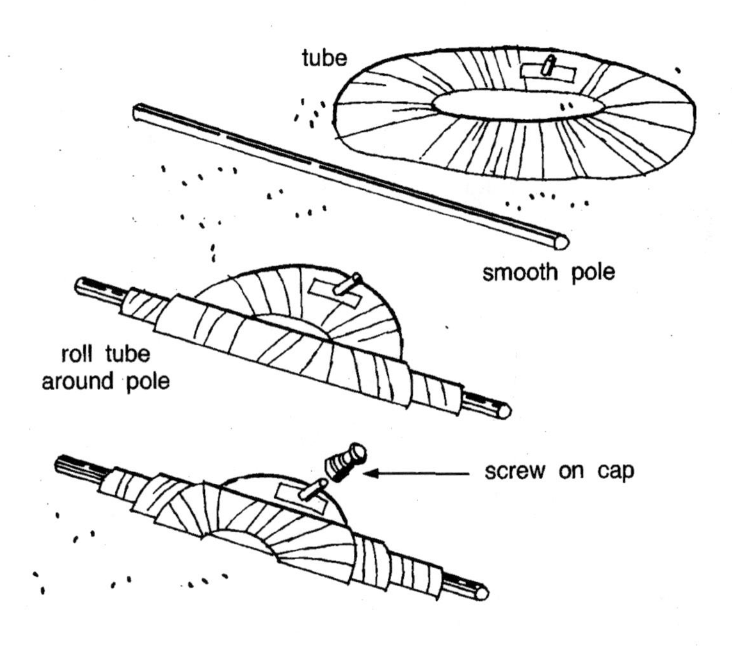

Note

When you screw the cap
on the air inlet,
do not put the valve back,
and keep the inner tube rolled up
until you attach the short gas line
(see Item 70 in this booklet).

65. Now you are ready
to attach the gas holder.

Attaching the gas holder

66. Cut a short piece from one end of the 12-metre gas line. This is to attach the inner tube to the pipe T-piece on the oil drum.

67. This piece should be long enough to connect the T-piece and the inner tube without being tight. If it is too tight, it may pull off.

68. Fold the short gas line at a place near the centre. Tie the fold tightly with cord. The drawing below will show you how.

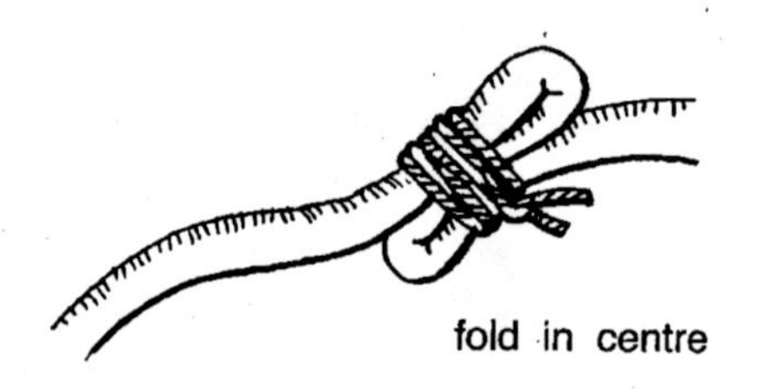

69. The fold will keep more air from getting into the inner tube when you attach it to the short piece of gas line.

70. Take the air inlet cap off the still-rolled inner tube and **attach the short gas line.** Be sure to attach it tightly. You may have to tie it with cord and seal it with tar or mastic. **Now you can unroll the inner tube** (see the drawings on the next page).

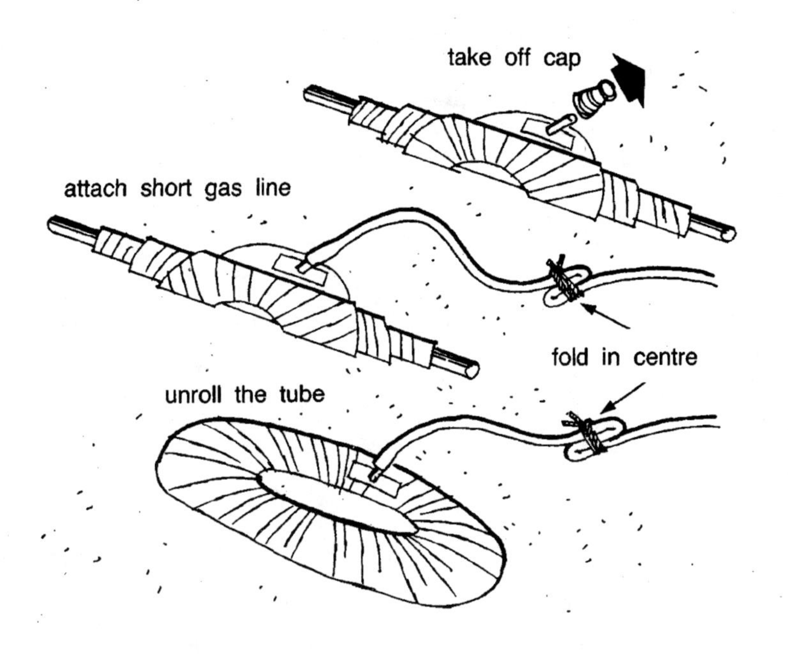

71. **It is very important to keep the gas holder from moving or the short gas line may pull off.**

72. If you are using a large inner tube, fit it over the oil drum and place it on the ground. The drum will keep the tube from moving.

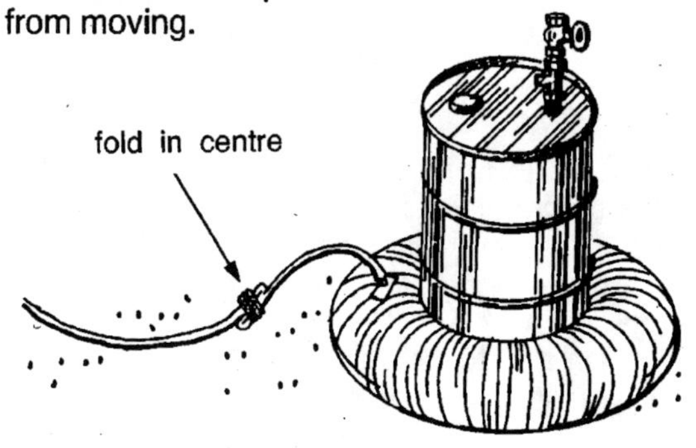

73. If you are using an inner tube that is too small to fit over the oil drum, you will have to keep it in place using wooden stakes.

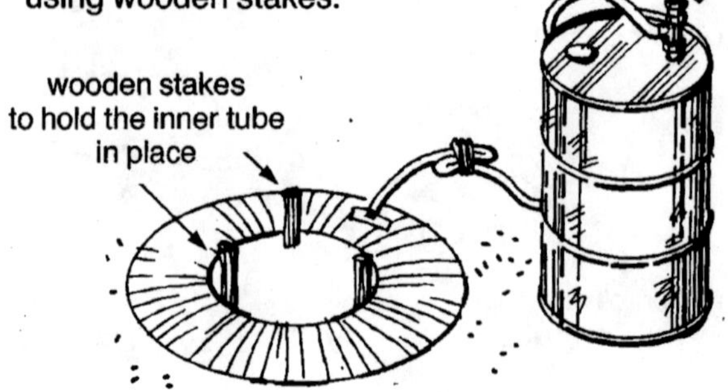

74. Connect the top of the short gas line to the pipe T-piece on the oil drum.

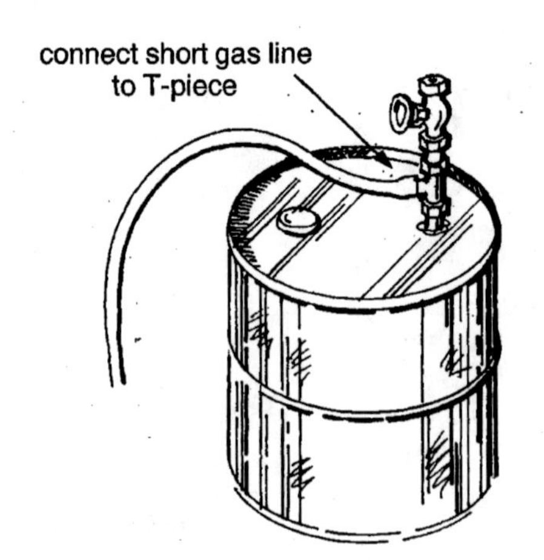

75. However, **do not** untie the fold in the centre of the short gas line (see Item 101 in this booklet) or attach the long gas line to the valve (see Items 89 to 122 in this booklet) until you are told to do so.

76. The drawings below
show you how to connect
both large and small inner tubes
to this kind of biogas unit.

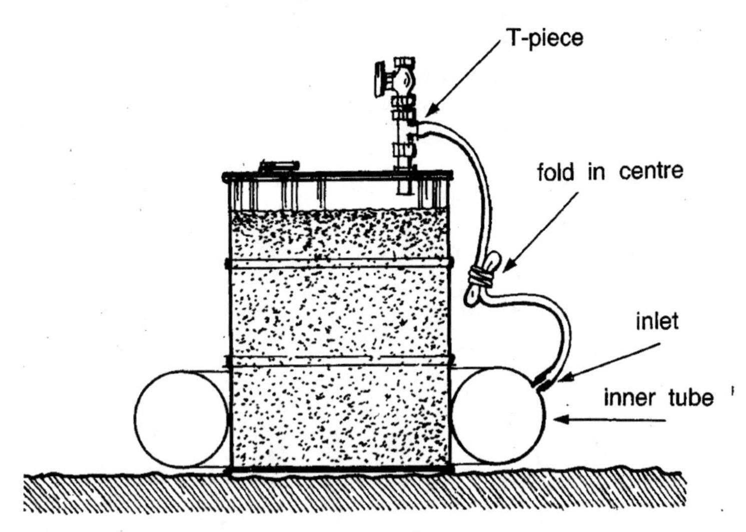

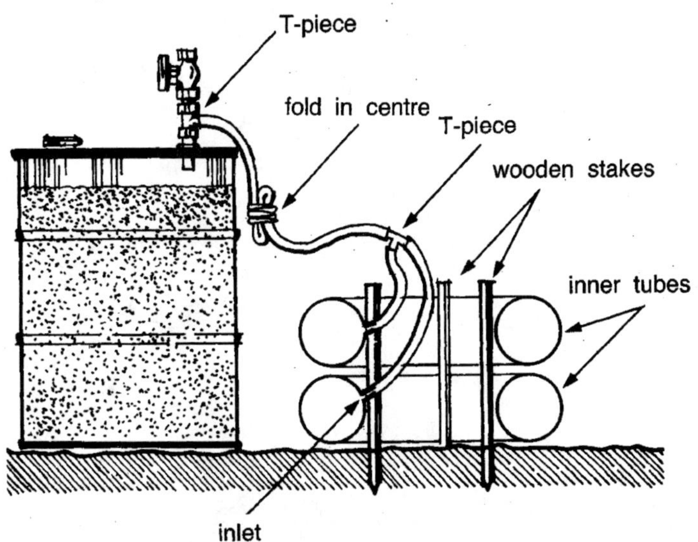

77. Now you are ready
to put waste into the oil drum.

PUTTING IN THE WASTE

The waste materials

78. You have already been told
that this biogas unit
is very much like the one
that you learned how to build
in Booklet No. 31.

79. Since your new biogas unit
is much the same as your old one,
you can use the same kind of wastes
in the same way.
Items 58 to 66 in Booklet No. 31
tell you how to prepare them.

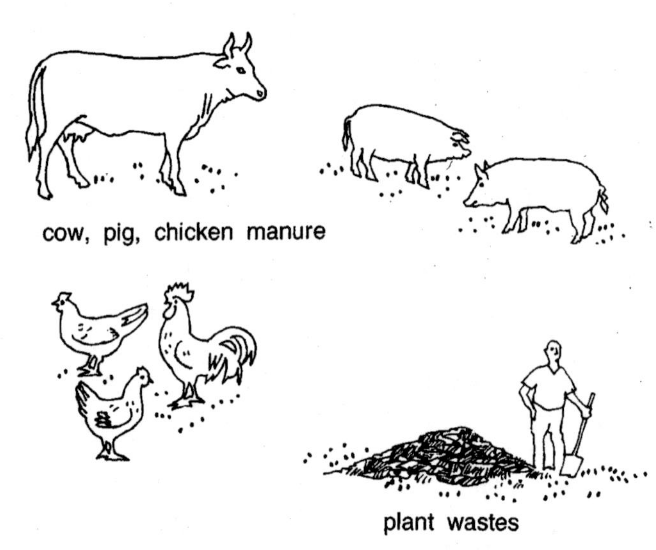

80. With the new unit,
as with your old unit,
you put in all of the waste
at **one** time when you begin.
Then, when **all** of the gas is made,
you take out **all** of the waste,
use it for fertilizer,
and begin all over again.

81. However, be especially careful
to mix the waste and water well.
Once this kind of unit is closed
you should not open it
until **all** of the gas is made.

82. You **cannot** stir it
or add more water
if the waste becomes too thick
as you could with your old unit
(see Items 94 to 96
in Booklet No. 31).

83. So, the waste and water mixture
for the new biogas unit
should be thin enough to pour easily.

84. It should be about as thin
as the paint or the whitewash
that you use to paint your house.

The starter

85. If your old unit is working,
take 4 litres of waste from it
to use as a starter when you begin.

86. However, if you do not have any waste
to use as a starter
you will have to make some.
Items 67 to 70 in Booklet No. 31
will tell you how to do it.

Putting waste in this biogas unit

87. When you are ready to add the waste, unscrew the plug in the waste hole and put it carefully aside. Put a large funnel in the hole.

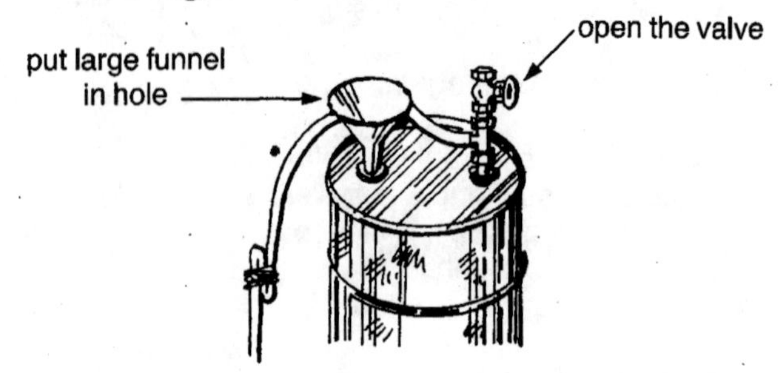

88. Open the valve so that when you add the waste the air that is inside the drum will be forced out through the gas outlet.

89. **You have not yet been told to attach the gas line and you should not have done so** (see Item 75 in this booklet).

90. Put three buckets of waste and three buckets of water in a large container and mix it well (see Item 84 in this booklet).

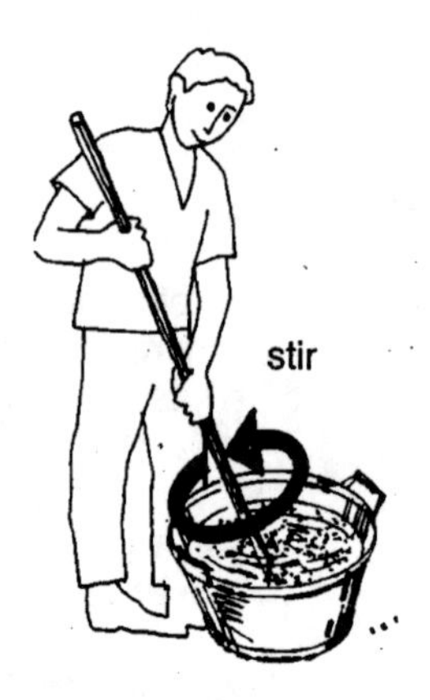

91. When the waste and water
are well mixed,
dip out a bucketful
and pour it through the funnel
into the oil drum.

92. If it does not flow
through the funnel,
add a **little** more water
to the waste mixture
in the large container.

93. Then try to pour
another bucketful through the funnel.
If the mixture is thin enough
to go through the funnel,
pour the rest into the drum.

94. Again put three buckets of waste
and three buckets of water
in the large container
and mix it as before.

95. Pour this mixture into the drum.
Then take out the funnel.
Put a pole long enough
to reach the bottom of the drum
into the waste hole
and stir all the mixture well.

96. Continue in this way
until the waste in the drum
is about 10 centimetres from the top.
Now put in about 4 litres of **starter**
and stir it well.

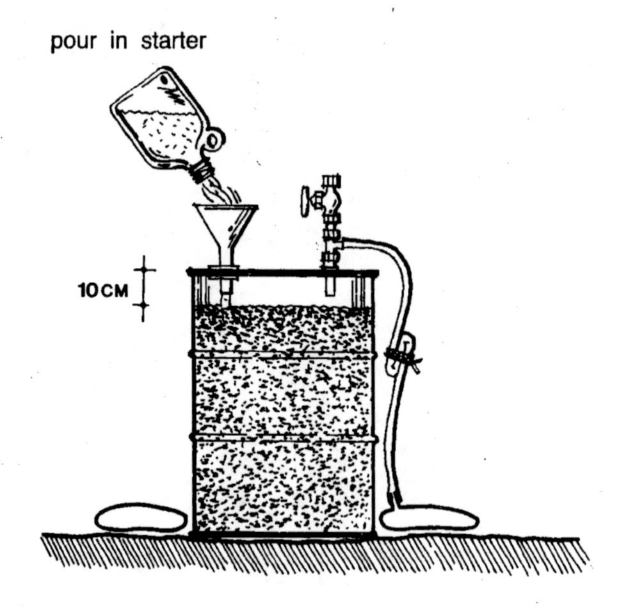

97. The starter,
which has already begun to work,
will help you to make gas sooner.

After the waste is in

98. Close the waste hole tightly and turn off the valve. After about two weeks, open the valve and let out all the gas that has collected in the top of the drum.

99. **While you are letting the gas out, be very careful not to have fire near the biogas unit.**

100. Listen as the gas escapes. When you hear the sound of the gas stop, turn off the valve **quickly**. This is to keep air from getting into the drum.

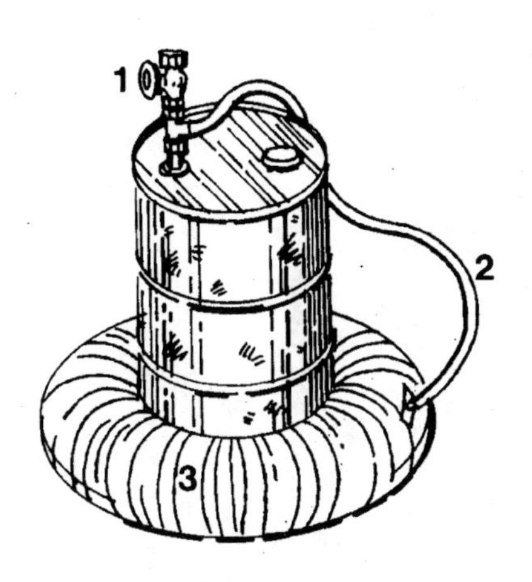

1 open the valve to let out the gas, then close it

2 untie fold in centre

3 when the tube begins to swell, gas is being made

101. **Now** you can untie the fold in the centre of the short gas line that runs to the inner tube gas holder. When you see the tube begin to swell, you will know that gas is being collected.

102. If you find that gas is leaking
from the top of the drum
after the unit has begun to work,
seal the leaks with tar,
mastic or paint
as you were told to do
in Item 47 in this booklet.

103. If gas is leaking
around the gas outlet, T-piece,
valve or inner tube, tighten them.

Note

A good way to check for leaks
after the biogas unit has begun to work
is to put soapy water on the drum
and on the joints of the parts and lines.
If you see bubbles anywhere
you will know that there is a leak.
Seal all leaks as you have been told to do.

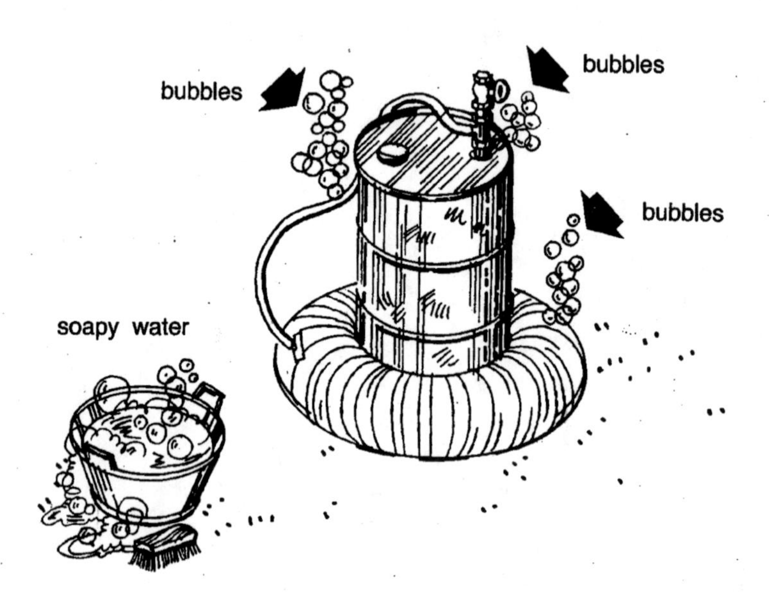

TIME

104. It may take up to three weeks
or even a month
for the waste in your new biogas unit
to begin making gas.
After that, the unit will continue
to make gas for about eight weeks.

105. During these eight weeks
half of the gas will be made
in the first two or three weeks
and the rest
in the last five or six weeks.

106. If you find that too little gas
is being made in the last weeks,
empty the unit and start again.

TEMPERATURE

107. You have been told in Booklet No. 31 that **biogas is best produced at a temperature between 32 and 37°C**. When the temperature is below 15°C, almost no gas is made.

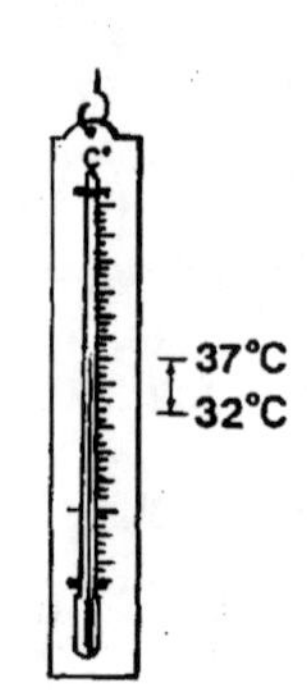

Cold weather protection

108. If the temperature where you live often falls below 15°C, you can keep the waste mixture warm by covering this biogas unit with plant materials such as leaves, grass, straw or maize stalks.

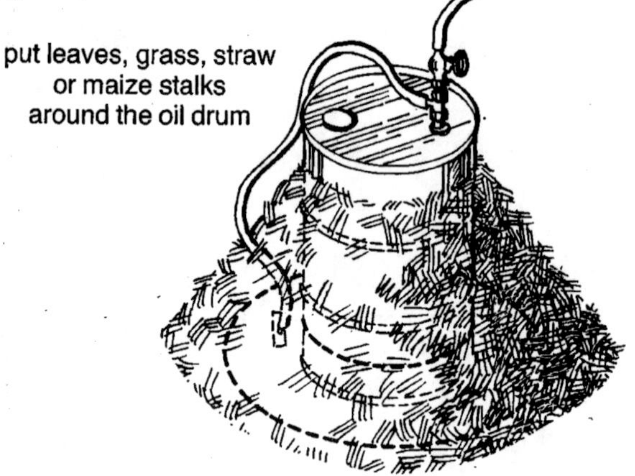

109. However, as you were told in Item 28 in this booklet, **you must not put this unit underground** as you could with your old unit or you will not be able to shake it to break up the scum

SCUM

110. Sometimes a hard layer of scum may form on top of the waste mixture in your biogas unit. If this happens, less gas will be made and gas will not collect in the inner tube.

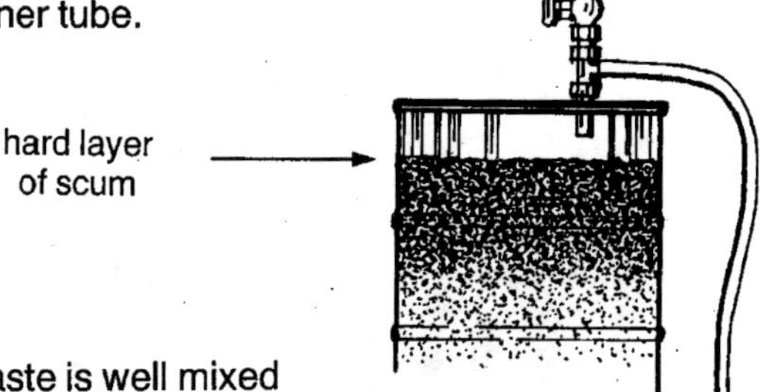

111. If the waste is well mixed before it is put into the unit, there will be less chance for scum to form and your biogas unit will make gas well.

112. Scum is more likely to form if you use plant materials than if you use only animal waste.

113. To keep scum from forming, shake your biogas unit from time to time. The drawing below shows you how.

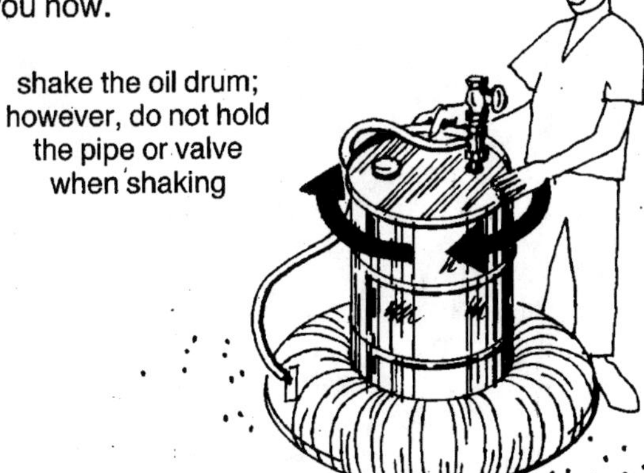

WHEN THE GAS IS MADE

114. **Do not burn
the first gas that is made
in your biogas unit.
It may have air in it
and could explode.**

115. A few days after the inner tube
has begun to swell with gas,
open the valve
and let out all of the gas
that has been collected.

116. **While you are letting the gas out
be very careful not to have fire
near the biogas unit.**

117. After the valve is open
you will have to force the gas
out of the inner tube or tubes.

118. You can force gas out of a tube
by rolling it as you were told to do
in Item 63 in this booklet,
or by putting a weight on it
such as pieces of wood or stones.

119. The drawings on the next page
show you how to force air
out of a biogas unit
with one or more inner tubes.

120. When all of the gas is out,
close the valve
and your biogas unit
will begin to collect gas again.

121. If you have done this carefully,
the next gas that is made
will have no air in it
and will be safe to burn.
Do not open the unit again
until all the gas has been made.

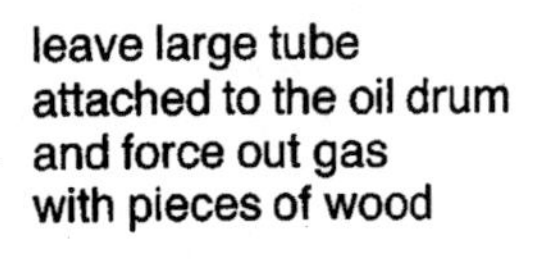

leave large tube
attached to the oil drum
and force out gas
with pieces of wood

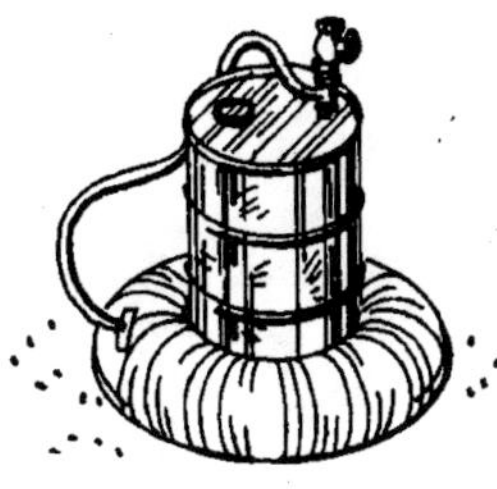

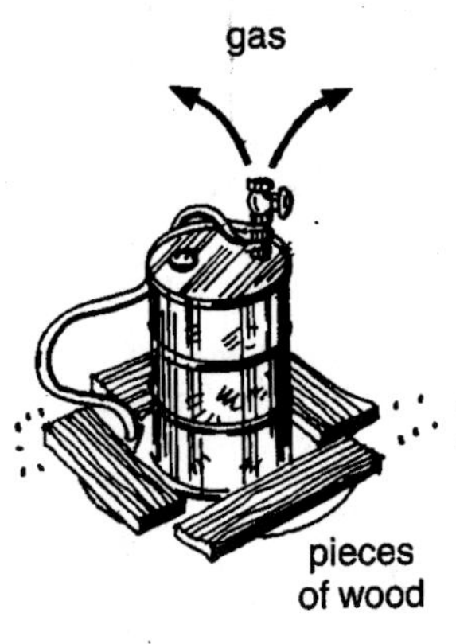

remove large tube
and force out gas
by rolling it

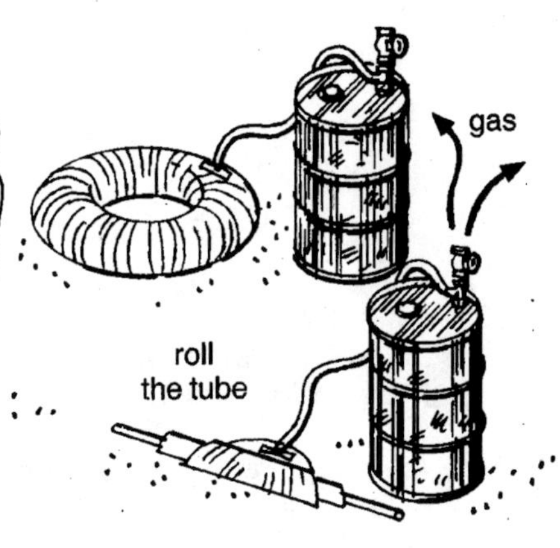

remove wooden stakes
and force out gas
with pieces of wood

122. **Now you can attach the gas line to the top of the valve.**
However, **do not** open the valve until the inner tube is half full.
Later, you can help to push the gas out of the inner tube by putting a few stones or bricks on it.

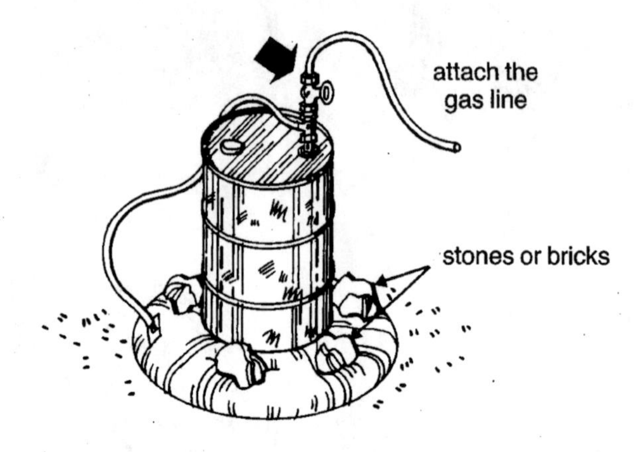

123. Items 108 to 114 in Booklet No. 31 tell you how to use biogas for cooking and how to clean the burner.

124. After all the gas has been made, take the unit apart and empty out the fertilizer.
Items 115 to 118 in Booklet No. 31 tell you how to use the fertilizer.

125. However, be sure to keep about 4 litres of the fertilizer to be used as a starter for the next time.

keep 4 litres of fertilizer as a starter

126. Clean the unit carefully
and check for leaks.

127. Now fill the unit
with new waste material
and add the starter.
Close the unit tightly
and it will begin to make gas again.

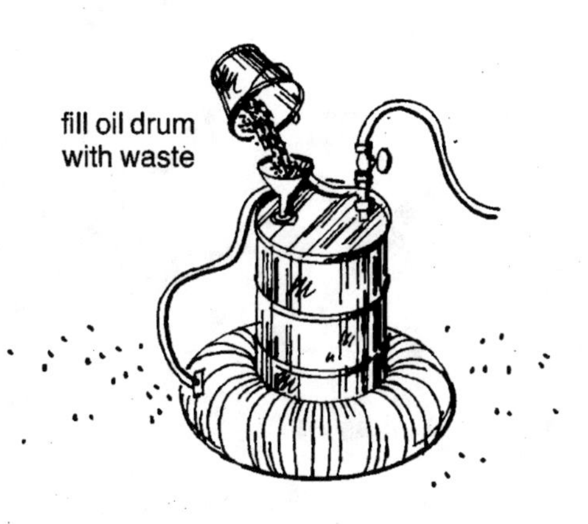

128. **Remember,
every time you start again,
do not burn
the first gas that is made.**

TAKING CARE OF YOUR BIOGAS UNIT

129. **Always be careful when you are near a biogas unit because gas may be leaking.**

130. **If gas is leaking and you breathe in too much of it, it can make you very sick.**

131. **Never build a fire, smoke, or even light a match near the unit, because if gas is leaking it may explode.**

132. Check your biogas unit and gas lines often to be sure that there are no leaks. The note on page 34 in this booklet tells you how to check for leaks in a working biogas unit.

133. If the oil drum begins to rust, coat it with the kind of paint that is used to paint metal.

134. About once each year, when you are taking the unit apart, wash it inside and outside with warm soapy water as you were told to do in Items 19 to 26 in this booklet.

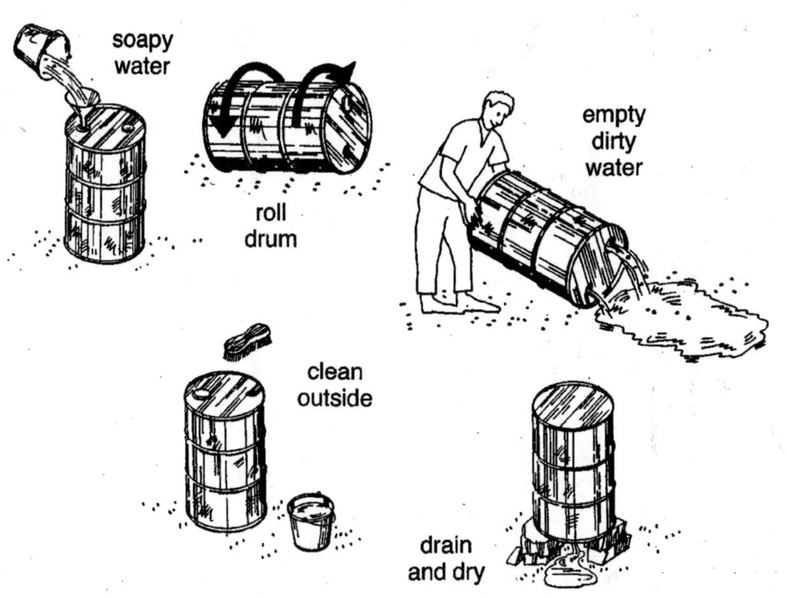

135. Then paint it inside and outside as you were told to do in Items 47 and 49 in this booklet.

MAKING MORE BIOGAS

136. As with your old biogas unit, the easiest way to make more gas is to build one or more small units and get gas from them all.

137. If you can get more oil drums, pipe, T-pieces, valves, inner tubes and gas lines, and if you have enough time, you can build and run more units.

138. The drawings below show you how to connect several units to the same gas line with T-pieces.

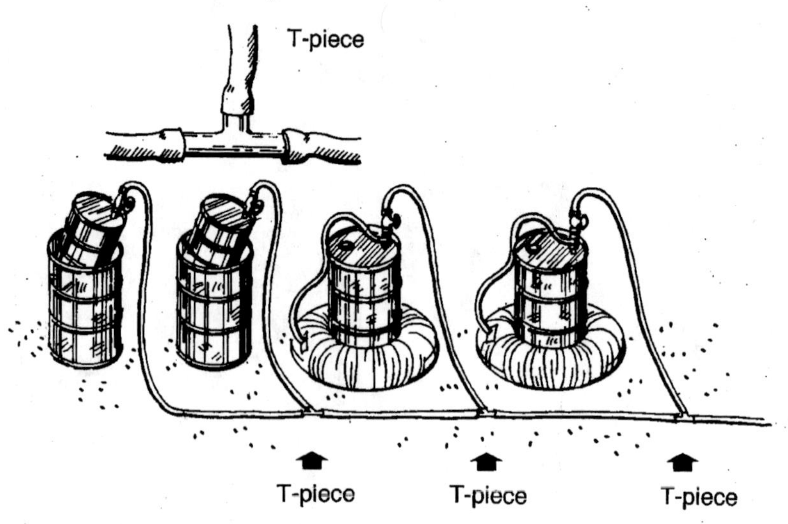

139. As you have already been told in Booklet No. 31, when you have several biogas units, fill them with waste at different times. That way, when all the gas in one unit has been used, you will get gas from another unit that is still working.

WHAT MORE CAN YOU DO?

140. The biogas unit
that you learned to build
in Booklet No. 31
and the biogas unit
that you learned to build
in this booklet
are both small units
that use one drum
for the waste holder.

141. With both of these small units
you put in all of the waste
when you first began.
Then, when all of the gas was made,
you cleaned out the unit,
used the waste as fertilizer
and started all over again.

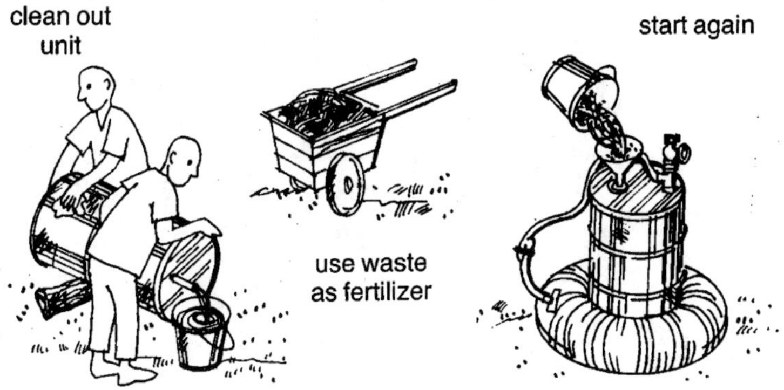

clean out unit

start again

use waste as fertilizer

142. By building and using
either or both of these units,
you learned a lot
from your experience.

143. Now, let us look
at still another biogas unit
that you can build
using what you have learned
to help you.

Another kind of biogas unit

144. This unit is bigger and better
than your first two units.
It is also more difficult
to build and to use.

145. This kind of biogas unit
can also be built using oil drums
with the same kind of pipe fittings
that you used before.

146. However, it can be built
much bigger than your old units.
You can use several oil drums
instead of only one.
So, you can make more gas
than you did before.

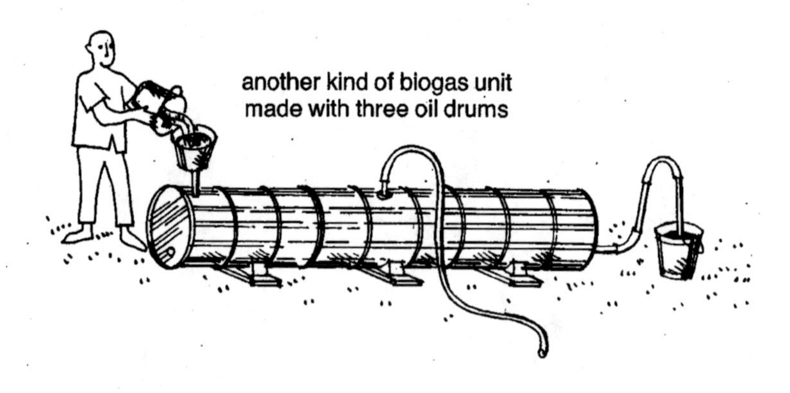

147. This kind of unit is
also filled with waste
when you first begin.

148. Then, after the unit
begins to make gas,
you continue to put in waste
from time to time.
You may do this every few days
or you may do it every week.

149. However, when you put new waste into this kind of unit, an equal amount of waste is pushed out of the unit.

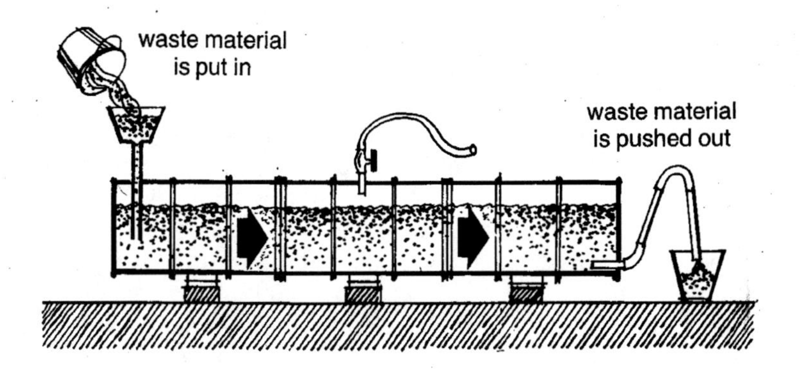

150. With fresh waste material always moving through the unit, it will make biogas for a much longer time.

151. You will learn more about this bigger and better unit in a later booklet in this series.

SAVE HARMLESS AGREEMENT

Because use of the information, instructions and materials discussed and shown in this book, document, electronic publication or other form of media is beyond our control, the purchaser or user agrees, without reservation to save Knowledge Publications Corporation, its agents, distributors, resellers, consultants, promoters, advisers or employees harmless from any and all claims, demands, actions, debts, liabilities, judgments, costs and attorney's fees arising out of, claimed on account of, or in any manner predicated upon loss of or damage to property, and injuries to or the death of any and all persons whatsoever, occurring in connection with or in any way incidental to or arising out of the purchase, sale, viewing, transfer of information and/or use of any and all property or information or in any manner caused or contributed to by the purchaser or the user or the viewer, their agents, servants, pets or employees, while in, upon, or about the sale or viewing or transfer of knowledge or use site on which the property is sold, offered for sale, or where the property or materials are used or viewed, discussed, communicated, or while going to or departing from such areas.

Laboratory work, scientific experiment, working with hydrogen, high temperatures, combustion gases as well as general chemistry with acids, bases and reactions and/or pressure vessels can be EXTREMELY DANGEROUS to use and possess or to be in the general vicinity of. To experiment with such methods and materials should be done ONLY by qualified and knowledgeable persons well versed and equipped to fabricate, handle, use and or store such materials. Inexperienced persons should first enlist the help of an experienced chemist, scientist or engineer before any activity thereof with such chemicals, methods and knowledge discussed in this media and other material distributed by KnowledgePublications Corporation or its agents. Be sure you know the laws, regulations and codes, local, county, state and federal, regarding the experimentation, construction and or use and storage of any equipment and or chemicals BEFORE you start. Safety must be practiced at all times. Users accept full responsibility and any and all liabilities associated in any way with the purchase and or use and viewing and communications of knowledge, information, methods and materials in this media.